Recent Titles in This Series

545 **John L. Lewis and Margaret A. M. Murray,** The method of layer potentials for the heat equation in time-varying domains, 1995

544 **Hans-Otto Walther,** The 2-dimensional attractor of $x'(t) = -\mu x(t) + f(x(t-1))$, 1995

543 **J. P. C. Greenlees and J. P. May,** Generalized Tate cohomology, 1995

542 **Alouf Jirari,** Second-order Sturm-Liouville difference equations and orthogonal polynomials, 1995

541 **Peter Cholak,** Automorphisms of the lattice of recursively enumerable sets, 1995

540 **Vladimir Ya. Lin and Yehuda Pinchover,** Manifolds with group actions and elliptic operators, 1994

539 **Lynne M. Butler,** Subgroup lattices and symmetric functions, 1994

538 **P. D. T. A. Elliott,** On the correlation of multiplicative and the sum of additive arithmetic functions, 1994

537 **I. V. Evstigneev and P. E. Greenwood,** Markov fields over countable partially ordered sets: Extrema and splitting, 1994

536 **George A. Hagedorn,** Molecular propagation through electron energy level crossings, 1994

535 **A. L. Levin and D. S. Lubinsky,** Christoffel functions and orthogonal polynomials for exponential weights on [-1,1], 1994

534 **Svante Janson,** Orthogonal decompositions and functional limit theorems for random graph statistics, 1994

533 **Rainer Buckdahn,** Anticipative Girsanov transformations and Skorohod stochastic differential equations, 1994

532 **Hans Plesner Jakobsen,** The full set of unitarizable highest weight modules of basic classical Lie superalgebras, 1994

531 **Alessandro Figà-Talamanca and Tim Steger,** Harmonic analysis for anisotropic random walks on homogeneous trees, 1994

530 **Y. S. Han and E. T. Sawyer,** Littlewood-Paley theory on spaces of homogeneous type and the classical function spaces, 1994

529 **Eric M. Friedlander and Barry Mazur,** Filtrations on the homology of algebraic varieties, 1994

528 **J. F. Jardine,** Higher spinor classes, 1994

527 **Giora Dula and Reinhard Schultz,** Diagram cohomology and isovariant homotopy theory, 1994

526 **Shiro Goto and Koji Nishida,** The Cohen-Macaulay and Gorenstein Rees algebras associated to filtrations, 1994

525 **Enrique Artal-Bartolo,** Forme de Jordan de la monodromie des singularités superisolées de surfaces, 1994

524 **Justin R. Smith,** Iterating the cobar construction, 1994

523 **Mark I. Freidlin and Alexander D. Wentzell,** Random perturbations of Hamiltonian systems, 1994

522 **Joel D. Pincus and Shaojie Zhou,** Principal currents for a pair of unitary operators, 1994

521 **K. R. Goodearl and E. S. Letzter,** Prime ideals in skew and q-skew polynomial rings, 1994

520 **Tom Ilmanen,** Elliptic regularization and partial regularity for motion by mean curvature, 1994

519 **William M. McGovern,** Completely prime maximal ideals and quantization, 1994

518 **René A. Carmona and S. A. Molchanov,** Parabolic Anderson problem and intermittency, 1994

(See the AMS catalog for earlier titles)

MEMOIRS
of the
American Mathematical Society

Number 545

The Method of Layer Potentials for the Heat Equation in Time-Varying Domains

John L. Lewis
Margaret A. M. Murray

March 1995 • Volume 114 • Number 545 (first of 4 numbers) • ISSN 0065-9266

American Mathematical Society
Providence, Rhode Island

1991 *Mathematics Subject Classification.*
Primary 42B20, 35K05.

Library of Congress Cataloging-in-Publication Data

Lewis, John L., 1943–
The method of layer potentials for the heat equation in time-varying domains / John L. Lewis, Margaret A. M. Murray.
p. cm. – (Memoirs of the American Mathematical Society, ISSN 0065-9266; no. 545)
"Volume 114, number 545 (first of 4 numbers)."
Includes bibliographical references.
ISBN 0-8218-0360-3 (alk. paper)
1. Heat equaiton. 2. Singular integrals. I. Murray, Margaret Anne Marie. II. Title. III. Series.
QA3.A57 no. 545
[QA377]
510 s–dc20
[515′.2433]

94-43211
CIP

Memoirs of the American Mathematical Society

This journal is devoted entirely to research in pure and applied mathematics.

Subscription information. The 1995 subscription begins with Number 541 and consists of six mailings, each containing one or more numbers. Subscription prices for 1995 are $369 list, $295 institutional member. A late charge of 10% of the subscription price will be imposed on orders received from nonmembers after January 1 of the subscription year. Subscribers outside the United States and India must pay a postage surcharge of $25; subscribers in India must pay a postage surcharge of $43. Expedited delivery to destinations in North America $30; elsewhere $92. Each number may be ordered separately; *please specify number* when ordering an individual number. For prices and titles of recently released numbers, see the New Publications sections of the *Notices of the American Mathematical Society.*

Back number information. For back issues see the *AMS Catalog of Publications.*

Subscriptions and orders should be addressed to the American Mathematical Society, P. O. Box 5904, Boston, MA 02206-5904. *All orders must be accompanied by payment.* Other correspondence should be addressed to Box 6248, Providence, RI 02940-6248.

Memoirs of the American Mathematical Society is published bimonthly (each volume consisting usually of more than one number) by the American Mathematical Society at 201 Charles Street, Providence, RI 02904-2213. Second-class postage paid at Providence, Rhode Island. Postmaster: Send address changes to Memoirs, American Mathematical Society, P. O. Box 6248, Providence, RI 02940-6248.

Printed in the United States of America.
This volume was printed directly from author-prepared copy.
♾ The paper used in this book is acid-free and falls within the guidelines established to ensure permanence and durability.
♻ Printed on recycled paper.

10 9 8 7 6 5 4 3 2 1 00 99 98 97 96 95

CONTENTS

ABSTRACT

This memoir consists of three papers (chapters I, II, III) in which we develop the method of layer potentials for the heat equation in time-varying domains. In chapter I we show certain singular integral operators on L^p are bounded. In chapter II we develop a modification of the David buildup scheme, as well as some extension theorems, to obtain L^p boundedness of the double layer heat potential on the boundary of our domains. In chapter III we use the results of the first two chapters, along with a buildup scheme, to show the mutual absolute continuity of parabolic measure and a certain projective Lebesgue measure. A_∞ results are also obtained. Moreover, we discuss the Dirichlet and Neumann problems for a certain subclass of our domains.

1991 *Mathematics Subject Classification.* Primary 42B20, 35K05.

keywords and phrases. heat equation, parabolic measure, Dirichlet problem, Neumann problem, layer potentials, time-varying domains, absolute continuity.

CHAPTER I
Singular Integrals

0. Introduction.

In recent years there has been renewed interest in the solution of parabolic boundary value problems by the method of layer potentials – a method which has been extraordinarily successful in the solution of elliptic problems. Building on the earlier work of Fabes and Rivère [FR], Russell Brown has used this method to solve Dirichlet and Neumann problems for the heat equation in Lipschitz cylinders, ([B1],[B2]). A natural next step is to develop the method of layer potentials for the heat equation in domains with time-dependent boundaries. This problem has been extensively studied in the case of a single space dimension (see [JT], [KW], [LeS], [LeMu1]). It has long been known that the boundary function must be at least $\mathrm{Lip}_{1/2}$ in time and Lip_1 in space, but this smoothness condition is not sufficient to permit the solution of the L^p Dirichlet problem by layer potential methods (see [KW]; see also [LeS], [LeMu 1,2]). A slightly smoother boundary, given by the graph of a function in a half-order BMO Sobolev space, turns out to be nearly optimal for the solution of the Dirichlet problem in the case of one space variable.

In higher dimensions, the heat equation, and indeed much more general parabolic equations and systems, have been extensively studied in bounded non-cylindrical domains. In the late 1950s and early 1960s Friedman([Fr1], [Fr2]), Solonnikov([So]), and others obtained *a priori* Schauder-type estimates for parabolic equations and systems in such domains. In the case of the heat equation, these estimates require that the boundary be given locally by the graph of a function in a parabolic Hölder class of order $2+\alpha$, i.e., the function, its spatial derivatives up to and including order 2, and its time derivative must satisfy a Lipschitz condition of order α. These estimates can then be used to prove existence-uniqueness theorems for the initial-Dirichlet problem for the heat equation when the data is in a Hölder class of order $2+\alpha$. Later work of Kemper([Ke]), in the early 1970s, used a boundary integral approach in noncylindrical domains that are regular(in the Perron-Wiener sense) for the Dirichlet problem, and obtained estimates for caloric measure depending on the $\mathrm{Lip}_{\frac{1}{2}} - \mathrm{Lip}_1$ norm of the boundary.

This chapter is the first of three chapters in which we develop the method of layer potentials for the heat equation in time-varying domains in R^{n+1}. Guided by our intuition in R^2, we shall assume that the boundary of the domain is given by the graph of a function which is Lip_1 in the space variables (uniformly in time) and which belongs to a half-order BMO Sobolev space relative to the time variable (uniformly in space). This mixed smoothness condition introduces a particular complexity into the problem of defining and estimating the layer potential operators. In particular, the function which describes the boundary is, in general, not differentiable with respect to time. In elliptic problems in Lipschitz domains, and in parabolic problems in Lipschitz cylinders (having time-independent boundaries),

[0]Research of both authors was supported in part by NSF grants.
Research of the first author was also supported in part by the Commonwealth of Kentucky through the Kentucky EPSCoR Program.
Received by the editor October 28, 1991, and in revised form September 20, 1993.

the Dirichlet problem, for example, is solved in terms of a double layer potential operator. The double layer potential is an integral operator defined by integration with respect to surface measure, whose kernel is the normal derivative of the fundamental solution on the boundary of the domain. In the graph domains which we are considering, surface measure on the boundary is not well-defined in the usual sense, and the normal derivative of the fundamental solution is a function of time. Threfore we introduce a kind of modified double layer potential operator, in which 'surface measure' on the boundary is replaced by the Lebesgue measure of the projection of the boundary graph onto R^n.

The modified double layer potential operator which we consider is a singular integral operator having mixed homogeneity. To obtain L^p estimates, it is natural to expand this operator into a series of multilinear operators, and then to study each of these operators using a mixed-homogeneity version of the T1 Theorem of David and Journé ([DJ]). Such a theorem was first proved by Lemarie ([L]), but we will use a simpler version based on ideas of Coifman and Meyer ([CM]). However, again the mixed conditions in space and time complicate matters considerably. In particular since our kernels need not be differentiable in the time variable, we cannot integrate by parts in this variable, as in [CM, p.13], to reduce a given multilinear singular integral operator to a simpler one. Also weak boundedness of our kernels is not obvious, since these kernels are not antisymmetric. In spite of these problems, Theorem 1 in this chapter shows that this program can be used to obtain L^p estimates for the above operators.

To be more specific we shall need some notation. Throughout this memoir we will let $x = (x_1, x_2, ..., x_{n-1})$ denote a point in R^{n-1} $(n > 1)$, and t will denote a point in R. For $\alpha \in (0,1)$, I_α is the Riesz potential operator of order α, defined for $g \in L^1_{\text{loc}}(R)$ by setting

$$I_\alpha(g)(t) = \int_R |s-t|^{\alpha-1} g(s)ds$$

(see [St]). The space of functions of bounded mean oscillation on R, BMO(R), is the space of functions $g \in L^1_{\text{loc}}(R)$ (modulo the constant functions) satisfying

$$||g||_* = \sup_I |I|^{-1} \int_I |g(t) - m_I(g)|dt < +\infty$$

where the supremum is taken over all subintervals I of R, $|I|$ is the Lebesgue measure of I, and $m_I(g)$ is the mean of g on I:

$$m_I(g) = |I|^{-1} \int_I g dt.$$

For the balance of the chapter, unless noted otherwise, we shall fix a continuous function $f: R^n \to R$, with compact support, for which there exist positive constants a_1, a_2 such that

$$|f(x,t) - f(y,t)| \le a_1|x-y|, \; x, y \in R^{n-1}, t \in R, \tag{0.1}$$

$$f(x,t) = I_\alpha(b(x,\cdot))(t), \; x, y \in R^{n-1}, \; t \in R, \tag{0.2}$$

where, for each fixed $x \in R^{n-1}$, $b(x,\cdot) \in \mathrm{BMO}(R)$ and

$$(0.3) \qquad ||b(x,\cdot)||_* \leq a_2.$$

From (0.2) and (0.3) it follows easily (as in [Str]) that

$$(0.4) \qquad |f(x,s) - f(x,t)| \leq ca_2|s-t|^\alpha, \ s,t \in R.$$

When $\alpha = 1/2$, we consider the graph domain given by

$$(0.5) \qquad D = \{(x,t) : x_n > f(x_1, x_2, ..., x_{n-1}, t)\}.$$

Recall that the fundamental solution to the heat equation in $R^{n+1} \setminus \{0\}$ is given by

(0.6)
$$W(x,t) = (4\pi t)^{-n/2} \exp\left(-\frac{|z|^2}{4t}\right) \chi_{[0,\infty)}(t), \ z \in R^n, \ t \in R, (z,t) \neq (0,0).$$

If D is a domain in $R^n \times R$, then for each fixed t we may regard $D \cap (R^n \times \{t\})$ as a domain in R^n, and when we do so we shall call it D_t. The outer unit normal to ∂D_t at a point $(x, f(x,t), t)$ is easily computed to be

$$\nu_t(x) = \frac{(\nabla_x f(x,t), -1)}{\sqrt{|\nabla_x f(x,t)|^2 + 1}}, \quad \nabla_x = \left(\frac{\partial}{\partial x_1}, \frac{\partial}{\partial x_2}, \dots, \frac{\partial}{\partial x_{n-1}}\right).$$

Associating in a natural way the point $(y,s) \in R^{n-1} \times R$ with the point $(y, f(y,s), s) \in \partial D$, we identify functions defined on $R^{n-1} \times R$ with functions defined on ∂D. With this in mind, we define the boundary double layer heat potential operator acting on a function $g : \partial D \to R$ by setting

(0.7)
$$Kg(x,t) = \tfrac{1}{2}(4\pi)^{-n/2} \int_{-\infty}^{t} \int_{R^{n-1}} \frac{\langle \nu_s(y), (x-y, f(x,t) - f(y,s))\rangle}{(t-s)^{n/2+1}}$$
$$\cdot \exp\left(\frac{-|x-y|^2 - (f(x,t) - f(y,s))^2}{4(t-s)}\right) \sqrt{1 + |\nabla_y f(y,s)|^2} g(y,s) dy ds.$$

The uniform Lipschitz estimate (0.1) assures that the partial derivatives of f with respect to the space variables are uniformly bounded in absolute value by a_1 (independent of time). Thus it is not difficult to see that L^p estimates for K may be deduced from L^p estimates for singular integrals of the form

(0.8)
$$\int_{-\infty}^{t} \int_{R^{n-1}} \frac{x_j - y_j}{(t-s)^{n/2+1}} \exp\left(\frac{-|x-y|^2 - (f(x,t) - f(y,s))^2}{4(t-s)}\right) g(y,s) dy ds$$

(0.9)
$$\int_{-\infty}^{t} \int_{R^{n-1}} \frac{f(x,t) - f(y,s)}{(t-s)^{n/2+1}} \exp\left(\frac{-|x-y|^2 - (f(x,t) - f(y,s))^2}{4(t-s)}\right) g(y,s) dy ds.$$

At least formally, the singular integrals (0.8) and (0.9) may be written as a series of multilinear expressions in f and g by expanding the exponential function in series. Specifically, we may rewrite (0.8) as the series

(0.10)

$$\sum_{m=0}^{\infty}\frac{(-1)^m}{4^m m!}\int_{-\infty}^{t}\int_{R^{n-1}}\frac{(x_j-y_j)[f(x,t)-f(y,s)]^{2m}}{(t-s)^{(2m+n)/2+1}}\exp\left(-\frac{|x-y|^2}{4(t-s)}\right)g(y,s)dyds,$$

while (0.9) has the formal series expansion

(0.11)

$$\sum_{m=0}^{\infty}\frac{(-1)^m}{4^m m!}\int_{-\infty}^{t}\int_{R^{n-1}}\frac{[f(x,t)-f(y,s)]^{2m+1}}{(t-s)^{(2m+n)/2+1}}\exp\left(-\frac{|x-y|^2}{4(t-s)}\right)g(y,s)dyds.$$

To estimate the terms appearing in these series, we consider a somewhat more general family of operators depending on the parameter $\alpha \in (0,1)$.

For every nonnegative integer m and for every $\epsilon > 0$, we define the family of truncated kernels $\Phi^m_{j,\epsilon}$, $1 \le j \le 2n$, in the following fashion. For $x, y \in R^{n-1}$ and $s, t \in R$ satisfying $|x-y| \ge \epsilon$, $|s-t| \ge \epsilon^{1/\alpha}$, we set

$$\Phi^m_{1,\epsilon}(x,t,y,s) = \frac{[f(y,s)-f(x,t)]^{2m+1}}{|s-t|^{(2m+n)\alpha+1}}\exp\left[-\frac{|x-y|^{1/\alpha}}{|s-t|}\right],$$

$$\Phi^m_{j,\epsilon}(x,t,y,s) = \frac{(x_{j-1}-y_{j-1})[f(y,s)-f(x,t)]^{2m}}{|s-t|^{2m+n)\alpha+1}}\exp\left[-\frac{|x-y|^{1/\alpha}}{|s-t|}\right], 2 \le j \le n,$$

$$\Phi^m_{j,\epsilon}(x,t,y,s) = \text{sgn}\,(s-t)\Phi^m_{j-n,\epsilon}(x,t,y,s)\exp\left[-\frac{|x-y|^{1/\alpha}}{|s-t|}\right],\quad n+1 \le j \le 2n.$$

If, on the other hand, $|x-y| < \epsilon$ or $|s-t| < \epsilon^{\frac{1}{\alpha}}$, then $\Phi^m_{j,\epsilon}(x,t,y,s) = 0$ for $1 \le j \le 2n$. The singular integral operators $\Lambda^m_{j,\epsilon}$ and Λ_j are defined, for $g \in C_0^\infty(R^{n-1} \times R)$, by setting

(0.12)

$$\begin{aligned}(\Lambda^m_j g)(x,t) &= \lim_{\epsilon\to 0}(\Lambda^m_{j,\epsilon}g)(x,t)\\ &= \lim_{\epsilon\to 0}\int_{R^{n-1}}\int_{R}\Phi^m_{j,\epsilon}(x,t,y,s)g(y,s)dyds\end{aligned}$$

for $1 \le j \le 2n$. The associated maximal operators $\bar{\Lambda}^m_j$ are defined by

(0.13)
$$\bar{\Lambda}^m_j g(x,t) = \sup_{\epsilon>0}|\Lambda^m_{j,\epsilon}g(x,t)|.$$

Our goal in the present chapter is to obtain L^p estimates for these operators, with precise control on the norm as a function of j, m, and the constants a_1 and a_2 appearing in (0.1) and (0.3). With this in view, we introduce the functions B_1, B_2 defined by

(0.14)

$$B_k(j,m,a_1,a_2) = \begin{cases}(a_k)^{2m-1}(a_1^2+a_2^2) & j=1 \text{ or } n+1;\ m\ge 1\\ (a_k)^{2m-2}(a_1^2+a_2^2) & 1\le j\le 2n;\ j\ne 1, n+1; m\ge 1\\ a_1+a_2 & m=0;\ j=1 \text{ or } n+1\\ 1 & 1\le j\le 2n;\ j\ne 1,n+1;\ m=0\end{cases}$$

for $k = 1, 2$. We prove

Theorem 1. *For $1 \leq j \leq 2n$, $1 < p < \infty$, the operator Λ_j^m may be extended to a bounded operator on $L^p(R^{n-1} \times R)$ so that for all $h \in L^p(R^{n-1} \times R)$,*

$$(\Lambda_j^m h)(x,t) = \lim_{\epsilon \to 0} (\Lambda_{j,\epsilon}^m h)(x,t) \tag{0.15}$$

exists for almost every $(x,t) \in R^{n-1} \times R$ relative to Lebesgue measure on R^n. Moreover, there exist positive constants $c(p)$ and c_1 such that, for all $h \in L^p(R^{n-1} \times R)$,

$$||\bar{\Lambda}_j^m h||_p \leq c(p)\beta(j,m,a_1,a_2,c_1)||h||_p \tag{0.16}$$

where

(0.17)
$$\beta(j,m,a_1,a_2,c_1) = c_1^{m+1}[\Gamma(2\alpha m + 1)B_1(j,m,a_1,a_2) + B_2(j,m,a_1,a_2)]$$

and Γ is the Euler gamma function.

The proof of Theorem 1 is in several steps. In section 1 of this chapter we prove the modified $T1$ theorem mentioned above. In section 2 we prove some basic estimates which will be used throughout the chapter. Using the groundwork laid in sections 1 and 2, we proceed in sections 3-5 to obtain L^2 estimates for a quite general class of operators. We are forced to take this rather roundabout approach, because of the difficulties mentioned earlier. Essentially we split the operator in Theorem 1, into a sum of operators, with the property that the "hard part " of each operator can be estimated using almost exclusively either (0.1) or (0.2)-(0.3) (and not both) . Thus Proposition 2.5 and Lemma 4.1 mainly use (0.2) and (0.3). They are generalizations of the main theorem in [Mu] . Furthermore, (0.1) is the key ingredient in the proof of Lemma 3.2 . Finally in section 5 we obtain in R^2 the L^2 estimate

$$||\Lambda_{j,\epsilon}^m h||_2 \leq \beta(j,m,a_1,a_2,c_1)||h||_2 \tag{0.18}$$

for $1 \leq j \leq 2n$ and $h \in L^2(R^2)$. The estimate (0.18) is extended to the case of general n by the method of rotations. Then we extend this theorem to $L^p(R^{n-1} \times R), 1 < p < \infty$.

Theorem 1 implies that the series in (0.10) and (0.11) converge in $L^p(R^n)$ for a_1 sufficiently small and $0 < a_2 < \infty$. In chapter 2 we shall use an argument modeled on the David buildup scheme and the technique in [LeMu2] to show that our double layer potential operator is bounded whenever $0 < a_1, a_2 < \infty$. Again the argument is complicated, because (0.1) and (0.2)-(0.3) do not mix well together. For example, (0.2) and (0.3) do not need to be preserved under a tilting of the graph of $x \to f(x,t)$, t fixed . In chapter 3, we will use the results in chapter 1 and chapter 2 to study the mutual absolute continuity of parabolic measure and a certain projective Lebesgue measure. We also investigate L^p Dirichlet and Neumann problems in graph domains, for small a_1, a_2.

A remark on notation: throughout the present chapter, 'c' denotes a positive constant depending only upon n and α, not necessarily the same at each occurrence.

More generally, '$c(\lambda_1, \lambda_2, ..., \lambda_k)$' will be used to denote a constant depending only upon n, α, and the variables $\lambda_1, \lambda_2, ..., \lambda_k$.

1. A T1 Theorem.

To simplify our analysis, we begin by estimating certain singular integral operators given by kernels having mixed homogeneity on $R \times R = R^2$, as a space of homogeneous type relative to the dilations

$$\delta_\lambda(x,t) = (\lambda^\alpha x, \lambda t). \tag{1.1}$$

When $\alpha = 1/2$, this is a family of parabolic dilations with respect to which the heat operator $\partial/\partial t - \partial^2/\partial x^2$ is homogeneous of degree one. For a thorough discussion of spaces of homogeneous type, the reader is referred to [A], [CW1], [CW2], and [L].

The T1 Theorem of David and Journé [DJ]) is among the most powerful tools available for the estimation of singular integrals; Lemarie ([L]) has extended it to the general setting of spaces of homogeneous type. We shall not need Lemarie's result in its full generality; instead, we state (and sketch the proof of) a simplified T1 Theorem adapted to the dilations (1.1) and to the special class of operators which are of interest to us here. Our result, and its proof, are inspired by the simplified proof of the David-Journé theorem due to Coifman and Meyer (see [CM]; see also the proof of Proposition 1 of [LeS]).

We begin with some notation and terminlogy. Relative to the dilations (1.1) we define the non-isotropic cube of "side length" $d > 0$ centered at $(x_0, t_0) \in R^2$ by

$$Q_d(x_0,t_0) = \{(x,t) \in R^2 : \ |x - x_0| < (d/2)^\alpha, \ |t - t_0| < d/2\}. \tag{1.2a}$$

$Q_d(x_0, t_0)$ is, in fact, a rectangle; its Lebesgue measure $|Q_d(x_0,t_0)|$ is equal to $2^{1-\alpha}d^{1+\alpha}$. It is more common in the literature to define a non-isotropic distance by setting

$$|||(x_0,t_0)||| = |x_0|^{1/\alpha} + |t| \tag{1.2b}$$

and then to work with non-isotropic balls of the form

$$B_d(x_0,t_0) = \{(x,t) \in R^2 : \ |||(x - x_0, t - t_0)||| < d\}; \tag{1.2c}$$

for our purposes, the non-isotropic cubes are easier to deal with. Relative to these cubes, we define the non-isotropic space $\mathrm{BMO}_\alpha = \mathrm{BMO}_\alpha(R^2)$ of functions of bounded mean oscillation as follows: a locally integrable function h on R^2 belongs to BMO_α provided that

$$||h||_{*,\alpha} = \sup_Q |Q|^{-1} \iint_Q |h(x,t) - m_Q(h)| dx dt < \infty. \tag{1.3}$$

Here, the supremum is taken over all non-isotropic cubes Q, and $m_Q(h)$ denotes the mean of h on Q:

$$m_Q(h) = |Q|^{-1} \iint_Q h(x,t) dx dt.$$

Given $\lambda > 0$, $(x_0, t_0) \in R^2$, and δ_λ as in (1.1), define the translate and dilate, $h_\lambda^{(x_0,t_0)}$, of a measurable function h by

$$\begin{aligned} h_\lambda^{(x_0,t_0)}(x,t) &= \lambda^{-1-\alpha} h(\delta_{\lambda^{-1}}(x_0 - x, t_0 - t)) \\ &= \lambda^{-1-\alpha} h\left(\tfrac{x_0-x}{\lambda^\alpha}, \tfrac{t_0-t}{\lambda}\right). \end{aligned}$$

The simple, non-isotropic dilate of h is given by

$$h_\lambda(x,t) = \lambda^{-1-\alpha} h(\delta_{\lambda^{-1}}(x,t)) = \lambda^{-1-\alpha} h\left(\frac{x}{\lambda^\alpha}, \frac{t}{\lambda}\right).$$

Since the Jacobian of $\delta_{\lambda^{-1}}$ is just $\lambda^{-1-\alpha}$, it is clear that when h is in L^1, so are $h_\lambda^{(x_0,t_0)}$ and h_λ, with the same L^1 norm.

We now describe the operators to which our T1 Theorem applies. Let H be a measurable function on R^4 such that

$$\text{(1.4)} \qquad \iint_{R^2} |H(x,t,y,s)| dyds, \quad \iint_{R^2} |H(y,s,x,t)| dyds < \infty$$

for almost every $(x,t) \in R^2$. For $g \in C_0^\infty(R^2)$, define

$$\text{(1.5)} \qquad \begin{aligned} Tg(x,t) &= \iint_{R^2} H(x,t,y,s) g(y,s) dyds, \\ T^* g(x,t) &= \iint_{R^2} H(y,s,x,t) g(y,s) dyds. \end{aligned}$$

We shall assume that T maps $C_0^\infty(R^2)$ boundedly into $L^2(R^2)$. We note that the integrability condition (1.4) permits us to sensibly define $T1$ and T^*1.

For $h, k \in L^2(R^2)$, we let

$$\langle h, k \rangle = \iint_{R^2} h(x,t) k(x,t) dxdt$$

denote the usual (real) inner product on $L^2(R^2)$. Let $\phi \in C_0^\infty(R)$ and $\theta \in C_0^\infty(R^2)$ be given by

$$\text{(1.6)} \qquad \begin{cases} \theta(x,t) = c\phi(x)\phi(t), \ (x,t) \in R^2, \\ \phi(u) = \chi_{(-1,1)}(u) \exp\left[-(1-u^2)^{-1}\right], \ u \in R. \end{cases}$$

Here c is a positive constant chosen so that $||\theta||_1 = 1$. When h is a function on R^2, we shall let Dh denote an arbitrary partial derivative of h with respect to either variable. With this notation and terminology, we now have the following result.

Special T1 Theorem. *Let T, T^* be as above. Suppose further that there exist constants $a_3, a_4, a_5 > 0$ and a nonnegative function $h \in L^1(R^2)$ satisfying*

$$\text{(1.7)} \qquad \iint_{R^2} (|y| + |s|^\alpha + 1)^\delta h(y,s) dyds = 1 \text{ for some } \delta \in (0,1),$$

such that

$$h(x_0,t_0) = h(|x_0|,|t_0|)\,, \tag{1.8}$$

$$|\langle (D\theta)_\lambda^{(x_0,t_0)}, T(\theta_\lambda^{(y_0,s_0)})\rangle| \le a_3 h_\lambda^{(y_0,s_0)}(x_0,t_0)\,, \tag{1.9}$$

$$|\langle (D\theta)_\lambda^{(x_0,t_0)},\ T^*(\theta_\lambda^{(y_0,s_0)})\rangle| \le a_4 h_\lambda^{(y_0,s_0)}(x_0,t_0)\,, \tag{1.10}$$

for all $x_0, y_0, s_0, t_0 \in R$, *and*

$$\max\{||\mathrm{T}1||_{*,\alpha}\,, ||T^*1||_{*,\alpha}\} \le a_5. \tag{1.11}$$

Then, for all $g \in C_0^\infty(R^2)$,

$$\max\{||Tg||_2\,, ||T^*g||_2\} \le c_5(a_3+a_4+a_5)||g||_2. \tag{1.12}$$

In order to sketch the proof of this special T1 Theorem, we shall require three lemmas which establish the interrelationships between BMO_α, Carleson measures, and Littlewood-Paley estimates for spaces of homogeneous type. These results are direct analogues of the corresponding statements in the usual, isotropic case; their proofs are straightforward and may be found in the dissertations of Aguirre and Lemarie ([A], [L]), so we shall omit them.

A positive Borel measure ν on $R^3_+ = R^2\times(0,\infty)$ is said to be a Carleson measure if and only if there is a constant C such that for all $(x_0,t_0)\in R^2$ and for all $d>0$,

$$\nu(Q_d(x_0,t_0)\times(0,d)) \le C|Q_d(x_0,t_0)|\,. \tag{1.13}$$

The smallest constants C for which (1.13) holds is called the Carleson constant for ν. We shall need three well known lemmas concerning Carleson measures and BMO_α. Recall that $\mathcal{S}(R^2)$ is the Schwartz class of C^∞ functions of rapid decrease on R^2 . Then

Lemma 1.1. *Let* $\psi\in\mathcal{S}(R^2)$ *satisfy* $\int_{R^2}\psi = 0$, *and suppose* $\beta\in BMO_\alpha$. *Then*

$$|(\psi_\lambda * \beta)(x,t)|^2\frac{dx\,dt\,d\lambda}{\lambda}$$

is a Carleson measure on R^3_+, *with Carleson constant less than or equal to* $C||\beta||^2_{*,\alpha}$. □

The second lemma is a well-known result whose proof relies on the interplay between nontangential maximal functions and Carleson measures.

Lemma 1.2. *Let* $\phi\in\mathcal{S}(R^2)$ *and suppose* ν *is a Carleson measure on* R^3_+ *with Carleson constant* C_ν. *Then for every* $p\in(1,\infty)$ *there is a constant* $C_{\phi,p}>0$ *such that for all* $k\in L^p(R^2)$,

$$\iiint_{R^3_+} |(\phi_\lambda * k)(x,t)|^p d\nu(x,t,\lambda) \le C_{\phi,p}C_\nu \iint_{R^2} |k(x,t)|^p dx\,dt. \ \square \tag{1.14}$$

Finally, we require a standard estimate for the non-isotropic Littlewood-Paley g-function:

Lemma 1.3. *Let ψ be as in Lemma 1.1. Then there is a constant $C_\psi > 0$ so that*

$$\iiint_{R^3_+} |(\psi_\lambda * k)(x,t)|^2 \frac{dx\,dt\,d\lambda}{\lambda} \leq C_\psi \iint_{R^2} |k(x,t)|^2 dx\,dt \tag{1.15}$$

for all $k \in L^2(R^2)$. □

Proof. With these lemmas in hand, we now sketch the proof of the special T1 Theorem. Our proof is modelled on the work of Coifman-Meyer in [CM], and on the proof of Proposition 1 in [LeS]. We write $P_\lambda g = \theta_\lambda * g$, so that

$$Tg = \lim_{\lambda \to 0} P_\lambda^2 T P_\lambda^2 g = \lim_{\epsilon \to 0} (T_\epsilon + T_\epsilon^*) g, \tag{1.16}$$

where

$$T_\epsilon g = -\int_\epsilon^{1/\epsilon} (\lambda \frac{\partial}{\partial \lambda} P_\lambda^2) T P_\lambda^2 g \frac{d\lambda}{\lambda}, \tag{1.17}$$

and T_ϵ^* is the adjoint of T_ϵ. Defining the Fourier transform of $k \in L^1(R^2)$ according to the normalization

$$\hat{k}(\xi, \tau) = \iint_{R^2} e^{-i(x\xi + t\tau)} k(x,t)\,dx\,dt$$

we have

$$\begin{aligned}
-\lambda \left(\frac{\partial}{\partial \lambda} P_\lambda^2 k \right)^\wedge (\xi, \tau) &= -\lambda \frac{\partial}{\partial \lambda} \left[\hat{\theta}^2(\lambda^\alpha \xi, \lambda \tau) \right] \hat{k}(\xi, \tau) \\
&= -2[\alpha \lambda^\alpha \xi \hat{\theta}(\lambda^\alpha \xi, \lambda\tau)(D_1 \hat{\theta})(\lambda^\alpha \xi, \lambda \tau) \\
&\quad + \lambda \tau \hat{\theta}(\lambda^\alpha \xi, \lambda \tau) D_2 \hat{\theta}(\lambda^\alpha \xi, \lambda \tau)] \hat{k}(\xi, \tau) \\
&= 2[\alpha (D_1 \theta)^\wedge (\lambda^\alpha \xi, \lambda \tau) \hat{\psi}_1(\lambda^\alpha \xi, \lambda \tau) \\
&\quad + (D_2 \theta)^\wedge (\lambda^\alpha \xi, \lambda \tau) \hat{\psi}_2(\lambda^\alpha \xi, \lambda \tau)] \hat{k}(\xi, \tau)
\end{aligned}$$

where D_j denotes differentiation with respect to the jth variable (for $j = 1, 2$) and $\psi_1(x,t) = x\theta(x,t)$, $\psi_2(x,t) = t\theta(x,t)$. If we let $R_{j,\lambda}$ denote convolution with $(D_j \theta)_\lambda$ and $Q_{j,\lambda}$ denote convolution with $(\psi_j)_\lambda$ for $j = 1, 2$, then

$$-\lambda \frac{\partial}{\partial \lambda} P_\lambda^2 = 2\alpha Q_{1,\lambda} R_{1,\lambda} + 2 Q_{2,\lambda} R_{2,\lambda}$$

so that

$$T_\epsilon = 2\alpha \int_\epsilon^{1/\epsilon} Q_{1,\lambda} L_{1,\lambda} P_\lambda g \frac{d\lambda}{\lambda} + 2\int_\epsilon^{1/\epsilon} Q_{2,\lambda} L_{2,\lambda} P_\lambda g \frac{d\lambda}{\lambda} \tag{1.18}$$

where $L_{j,\lambda} = R_{j,\lambda} T P_\lambda$, $j = 1, 2$. It is easily seen that

$$L_{j,\lambda} k(x,t) = \iint_{R^2} l_{j,\lambda}(x,t,y,s) k(y,s) dyds$$

where

$$l_{j,\lambda}(x,t) = \langle (D_j\theta)_\lambda^{(x,t)},\ T(\theta_\lambda^{(y,s)})\rangle.$$

Moreover, $L_{j,\lambda}1$ is easily seen to equal $R_{j,\lambda}\beta$, where $\beta = \mathrm{T}1$ (to see this, use the fact that $\hat{\theta}(0) = 1$). So we may write

$$\int_\epsilon^{1/\epsilon} Q_{j,\lambda} L_{j,\lambda} P_\lambda g \frac{d\lambda}{\lambda} = \int_\epsilon^{1/\epsilon} Q_{j,\lambda}[(R_{j,\lambda}\beta)(P_\lambda g)]\frac{d\lambda}{\lambda} + E_\epsilon(g) \tag{1.19}$$

where the "error term" $E_\epsilon(g)$ is given by

$$\int_\epsilon^{1/\epsilon} Q_{j,\lambda}\{(L_{j,\lambda}P_\lambda)g - (L_{j,\lambda}1)(P_\lambda g)\}\frac{d\lambda}{\lambda}. \tag{1.20}$$

The first term on the right-hand side of (1.19) is easily estimated by duality using the lemmas. Indeed, let $k \in L^2(R^2)$; then

$$|\langle k, \int_\epsilon^{1/\epsilon} Q_{j,\lambda}[|(R_{j,\lambda}\beta)(P_\lambda g)]\frac{d\lambda}{\lambda}\rangle|$$

$$\leq \left(\iiint_{R^3_+} |Q_{j,\lambda} k(x,t)|^2 \frac{dxdtd\lambda}{\lambda} \right)^{1/2} \tag{1.21}$$

$$\cdot \left(\iiint_{R^3_+} |R_{j,\lambda}\beta(x,t)|^2 |P_\lambda g(x,t)|^2 \frac{dxdtd\lambda}{\lambda} \right)^{1/2}$$

$$\leq c||\beta||_{*,\alpha}||g||_2||k||_2$$

by Lemma 1.1 with $\psi = D_j\theta$, Lemma 1.2 with $\phi = \theta$, and Lemma 1.3 with $\psi = \psi_j$. By (1.11) and (1.21) we have

$$|| \int_\epsilon^{1/\epsilon} Q_{j,\lambda}[(R_{j,\lambda}\beta)(P_\lambda g)]\frac{d\lambda}{\lambda}||_2 \leq ca_5||g||_2. \tag{1.22}$$

We now turn our attention to $E_\epsilon(g)$, which we shall also estimate by duality. For $k \in L^2(R^2)$,

(1.23)

$$|\langle k, E_\epsilon(g)\rangle| = |\int_\epsilon^{1/\epsilon} \iint_{R^2} (Q_{j,\lambda}k)\{(L_{j,\lambda}P_\lambda)g - (L_{j,\lambda}1)(P_\lambda g)\}(x,t)dxdt\frac{d\lambda}{\lambda}|$$

$$\leq \left(\iiint_{R^3_+} |Q_{j,\lambda}k(x,t)|^2 \frac{dxdtd\lambda}{\lambda}\right)^{1/2}$$

$$\cdot \left\{ \int_\epsilon^{1/\epsilon} \iint_{R^2} \left| \iint_{R^2} l_{j,\lambda}(x,t,y,s)[(P_\lambda g)(y,s) - (P_\lambda g)(x,t)]dyds \right|^2 \frac{dxdtd\lambda}{\lambda}\right\}^{1/2}$$

$$\leq ca_3||k||_2 \left\{ \int_\epsilon^{1/\epsilon} \iint_{R^2} [\iint_{R^2} h_\lambda^{(y,s)}(x,t)|(P_\lambda g)(y,s) - (P_\lambda g)(x,t)|^2 dyds] \frac{dxdtd\lambda}{\lambda}\right\}^{\frac{1}{2}}$$

where we have used Lemma 1.3 with $\psi = \psi_j$, the estimate (1.9), the Schwarz inequality, and the fact that h is nonnegative with $||h||_1 \leq 1$. Now note that

(1.24)

$$\iint_{R^2} \left[\iint_{R^2} h_\lambda^{(y,s)}(x,t)|(P_\lambda g)(y,s) - (P_\lambda g)(x,t)|^2 dyds \right] dxdt$$

$$= \iint_{R^2} \left[\iint_{R^2} h_\lambda(u,v)|(P_\lambda g)(x+u, t+v) - (P_\lambda g)(x,t)|^2 dudv \right] dxdt$$

$$= \iint_{R^2} \left[\iint_{R^2} h_\lambda(u,v)|e^{i(\xi\cdot u + \tau\cdot v)} - 1|^2 |\hat{\theta}(\lambda^\alpha \xi, \lambda\tau)|^2 |\hat{g}(\xi,\tau)|^2 dudv \right] d\xi d\tau$$

by Plancherel's Theorem and Fubini's Theorem. Now there exists a constant $C_\delta > 0$ such that

$$|e^{i(\xi\cdot u+\tau\cdot v)} - 1|^2 \leq C_\delta \left(\left|\frac{u}{\lambda^\alpha}\right| + \left|\frac{v}{\lambda}\right|^\alpha + 1\right)^\delta (|\lambda^\alpha\xi| + |\lambda\tau|^\alpha)^\delta. \tag{1.25}$$

Making use of (1.7), (1.25) and a straightforward change of variables, we see that

(1.24) is majorized by

$$C_\delta \iint_{R^2} (|\lambda^\alpha \xi| + |\lambda \tau|^\alpha)^\delta |\hat{\theta}(\lambda^\alpha \xi, \lambda \tau)|^2 |\hat{g}(\xi, \tau)|^2 d\xi d\tau. \tag{1.26}$$

Thus, by (1.23), (1.26), and Fubini,

(1.27)

$$|\langle k, E_\epsilon(g)\rangle| \leq C_\delta a_3 ||k||_2 \left[\iint_{R^2} \left\{ \int_\epsilon^{1/\epsilon} (|\lambda^\alpha \xi| + |\lambda \tau|^\alpha)^\delta |\hat{\theta}(\lambda^\alpha \xi, \lambda \tau)|^2 \frac{d\lambda}{\lambda} \right\} |\hat{g}(\xi, \tau)|^2 d\xi d\tau \right]^{1/2}.$$

If $(\xi, \tau) \in R^2$ and $\mu = |||(\xi, \tau)|||$ is its non-isotropic distance from the origin as given by (1.2b), then there exists (ξ', τ') with $|||(\xi', \tau')||| = 1$ such that $(\xi, \tau) = \delta_\mu(\xi', \tau')$. Under the change of variables $s = \mu\lambda$, the inner integral in (1.27) becomes

$$\begin{aligned} &\int_{\mu\epsilon}^{\mu/\epsilon} (|s^\alpha \xi'| + |s\tau'|^\alpha)^\delta |\hat{\theta}(s^\alpha \xi', s\tau')|^2 \frac{ds}{s} \\ &\leq \int_0^\infty (|s^\alpha \xi'| + |s\tau'|^\alpha)^\delta |\hat{\theta}(s^\alpha \xi', s\tau')|^2 \frac{ds}{s}, \end{aligned} \tag{1.28}$$

and this last expression defines a continuous function on the non-isotropic 'unit sphere'

$$\partial B_1(0,0) = \{(\xi', \tau') \in R^2 : \; |||(\xi', \tau')||| = 1\}$$

(see (1.2c)). Since the sphere is compact, (1.28) is majorized by a constant depending on α and δ. From this we see that

$$|\langle k, E_\epsilon(g)\rangle| \leq C_\delta ||k||_2 ||g||_2 a_3 \tag{1.29}$$

by an application of Plancherel's Theorem. Combining (1.18), (1.19), (1.22), and (1.29), we have

$$||T_\epsilon g||_2 \leq C_\delta (a_3 + a_5) ||g||_2. \tag{1.30}$$

A similar analysis applies to T_ϵ^*, so that the estimate (1.12) follows from (1.16). □

2. Basic Estimates.

Throughout this section, we let n be a fixed integer greater than 1. In preparation for our first application of the T1 Theorem, we introduce the family of functions $V = V_{i,l}$, indexed by nonnegative integers i and l, on R^2. For each choice of i, l set $\sigma = \sigma(i,l) = (l + n + i + 1)\alpha + 1$ and let $d\xi = d\xi_1 d\xi_2 \ldots d\xi_{i-1} d\xi_i$; then we define

$$V_{0,l}(z,u) = |z|^{n-2+l} |u|^{-\sigma} \exp(-|z|^{1/\alpha} |u|^{-1}), \tag{2.1 a}$$

$$V_{1,l}(z,u) = |z|^l |u|^{-\sigma} \int_{|z|}^\infty \xi_1^{n-2} \exp(-\xi_1^{1/\alpha} |u|^{-1}) d\xi_1, \tag{2.1b}$$

$$(2.1c)\quad V_{i,l}(z,u)=|z|^l|u|^{-\sigma}\int_{|z|}^{\infty}\int_{\xi_i}^{\infty}\cdots\int_{\xi_3}^{\infty}\int_{\xi_2}^{\infty}\xi_1^{n-2}\exp(-\xi_1^{1/\alpha}|u|^{-1})d\xi.$$

for $i\geq 2$. We also introduce the family of functions $W=W_{i,l}$, indexed by a positive integer i and a nonnegative integer l:

$$(2.2a)\qquad W_{1,l}(z,u)=|z|^l|u|^{-\sigma-2}\int_{|z|}^{\infty}\xi_1^{n-2+1/\alpha}\exp\left(-\frac{\xi_1^{1/\alpha}}{|u}\right)d\xi_1,$$

$$(2.2b)\quad W_{i,l}(z,u)=|z|^l|u|^{-\sigma-2}\int_{|z|}^{\infty}\int_{\xi_i}^{\infty}\cdots\int_{\xi_2}^{\infty}\xi_1^{n-2+1/\alpha}\exp\left(-\frac{\xi_1^{1/\alpha}}{|u|}\right)d\xi.$$

We begin with some simple calculations which are basic to all that follows.

Lemma 2.1. *Let i and l be nonnegative integers. Then we have*

$$(2.3)\qquad \int_0^\infty |z|^\lambda|u|^\nu V_{i,l}(z,u)dz = \frac{\alpha\Gamma(\alpha(n+l+\lambda+i-1))\Gamma(l+\lambda+1)}{\Gamma(l+\lambda+i+1)}|u|^{\nu+(\lambda-2)\alpha-1}$$

for all $\lambda>1-(n+l+i)$ and for all real ν; and

$$(2.4)\qquad \int_0^\infty |z|^\lambda u^\nu V_{i,l}(z,u)du = \frac{\Gamma(\alpha(n+l+i+1)-\nu)\Gamma(l+3-\nu/\alpha)}{\Gamma(l+i+3-\nu/\alpha)}|z|^{\lambda-3+\nu/\alpha}$$

for all $\lambda\in R$ and for all $\nu<(l+3)\alpha$. Moreover, for $i\geq 1$ we also have

$$(2.5)\qquad \int_0^\infty |z|^\lambda|u|^\nu W_{i,l}(z,u)dz = \frac{\alpha\Gamma(l+\lambda+1)\Gamma(\alpha(n+l+i+\lambda-1)+1)}{\Gamma(l+\lambda+i+1)}|u|^{\nu+(\lambda-2)\alpha-2}$$

for all $\lambda>1-(n+l+i+1/\alpha)$ and for all real ν, and

$$(2.6)\qquad \int_0^\infty |z|^\lambda u^\nu W_{i,l}(z,u)du = \frac{\Gamma(\alpha(n+l+1)+2-\nu)\Gamma(l+3+\frac{1-\nu}{\alpha})}{\Gamma(l+i+3+\frac{1-\nu}{\alpha})}|z|^{\lambda-3+\frac{\nu-1}{\alpha}}$$

for all $\lambda\in R$ and for all $\nu<(l+3)\alpha+1$.

Proof. We begin by observing that whenever $\delta>-1$, we have

$$(2.7)\qquad \begin{aligned}\int_0^\infty z^\delta\exp(-z^{1/\alpha}|u|^{-1})dz &= \alpha|u|^{(\delta+1)\alpha}\int_0^\infty t^{(\delta+1)\alpha-1}e^{-t}dt\\ &= \alpha|u|^{(\delta+1)\alpha}\Gamma((\delta+1)\alpha),\end{aligned}$$

upon making the change of variables $t = z^{1/\alpha}|u|^{-1}$; and for $\delta > 1$, we have

$$\int_0^\infty u^{-\delta}\exp\left(-|z|^{1/\alpha}u^{-1}\right)du = |z|^{-(\delta-1)/\alpha}\int_0^\infty t^{\delta-2}e^{-t}dt \tag{2.8}$$
$$= |z|^{-(\delta-1)/\alpha}\Gamma(\delta-1)$$

upon making the change of variables $t = |z|^{1/\alpha}u^{-1}$. When $i = 0$, taking $\delta = n+l+\lambda-2$ in (2.7) yields

$$\begin{aligned}\int_0^\infty z^\lambda|u|^\nu V(z,u)dz &= |u|^{\nu-\sigma}\int_0^\infty z^{n+l+\lambda-2}\exp\left(-z^{1/\alpha}|u|^{-1}\right)dz\\ &= \alpha|u|^{\nu-\sigma}|u|^{(n+l+\lambda-1)\alpha}\Gamma(\alpha(n+l+\lambda-1))\\ &= \alpha|u|^{\nu+(\lambda-2)\alpha-1}\Gamma(\alpha(n+l+\lambda-1)).\end{aligned}$$

For $i = 1$, we have

$$\begin{aligned}\int_0^\infty z^\lambda|u|^\nu V(z,u)dz &= |u|^{\nu-\sigma}\int_0^\infty z^{\lambda+l}\int_z^\infty \xi_1^{n-2}\exp\left(-\xi_1^{1/\alpha}|u|^{-1}\right)d\xi_1 dz\\ &= |u|^{\nu-\sigma}\int_0^\infty\left(\int_0^{\xi_1} z^{\lambda+l}dz\right)\xi_1^{n-2}\exp\left(-\xi_1^{1/\alpha}|u|^{-1}\right)d\xi_1\\ &= \frac{1}{\lambda+l+1}|u|^{\nu-\sigma}\int_0^\infty \xi_1^{n+l+\lambda-1}\exp\left(-\xi_1^{1/\alpha}|u|^{-1}\right)d\xi_1\\ &= \frac{\alpha}{\lambda+l+1}|u|^{\nu-\sigma}|u|^{(n+l+\lambda)\alpha}\Gamma(\alpha(n+l+\lambda))\\ &= \frac{\alpha\Gamma(\alpha(n+l+\lambda))\Gamma(l+\lambda+1)}{\Gamma(l+\lambda+2)}|u|^{\nu+(\lambda-2)\alpha-1}\end{aligned}$$

by Fubini's theorem and (2.7) with $\delta = n+l+\lambda-1$. Finally, for $i \geq 2$, we have

$$\begin{aligned}&\int_0^\infty z^\lambda|u|^\nu V(z,u)dz\\ &= |u|^{\nu-\sigma}\int_0^\infty z^{\lambda+l}\int_z^\infty\int_{\xi_i}^\infty\cdots\int_{\xi_2}^\infty \xi_1^{n-2}\exp\left(-\xi_1^{1/\alpha}|u|^{-1}\right)d\xi dz\\ &= |u|^{\nu-\sigma}\int_0^\infty \xi_1^{n-2}\exp\left(-\frac{\xi_1^{1/\alpha}}{|u|}\right)\left[\int_0^{\xi_1}\int_0^{\xi_2}\cdots\int_0^{\xi_{i-1}}\int_0^{\xi_i} z^{\lambda+l}dzd\xi_i\cdots d\xi_3 d\xi_2\right]d\xi_1;\end{aligned}$$

the integral in brackets is easily seen to equal

$$\frac{1}{(\lambda+l+1)(\lambda+l+2)\cdots(\lambda+l+i)}\xi_1^{\lambda+l+i} = \frac{\Gamma(l+\lambda+1)}{\Gamma(l+\lambda+1+i)}\xi_1^{\lambda+l+i},$$

so that, taking $\delta = n+l+\lambda+i-2$ in (2.7) yields

$$\int_0^\infty z^\lambda |u|^\nu V(z,u)dz$$

$$= \frac{\alpha\Gamma(\alpha(n+l+\lambda+i-1))\Gamma(l+\lambda+1)}{\Gamma(l+\lambda+i+1)}|u|^{\nu-\sigma}|u|^{\alpha(n+l+\lambda+i-1)}$$

$$= \frac{\alpha\Gamma(\alpha(n+l+\lambda+i-1))\Gamma(l+\lambda+1)}{\Gamma(l+\lambda+i+1)}|u|^{\nu+(\lambda-2)\alpha-1}.$$

This establishes (2.3) in all cases; the restriction on λ in each case arises from the constant $\delta > -1$ in (2.7). When $i = 0$, we have

$$\begin{aligned}\int_0^\infty |z|^\lambda u^\nu V(z,u)du &= |z|^{n+l+\lambda-2}\int_0^\infty u^{\nu-\sigma}\exp(-|z|^{\frac{1}{\alpha}}u^{-1})du \\ &= |z|^{\lambda-3+\nu/\alpha}\,\Gamma(\alpha(n+l+1)-\nu),\end{aligned}$$

taking $\delta = \sigma-\nu = \alpha(n+l+1)-\nu+1$ in (2.8) ; the equality here is valid for all ν satisfying $\nu < \alpha(n+l+1)$. When $i = 1$, we have

$$\int_0^\infty |z|^\lambda u^\nu V(z,u)du$$

$$= |z|^{\lambda+l}\int_0^\infty u^{\nu-\sigma}\int_{|z|}^\infty \xi_1^{n-2}\exp\left(-\xi_1^{1/\alpha}u^{-1}\right)d\xi_1 du$$

$$= |z|^{\lambda+l}\int_{|z|}^\infty \xi_1^{n-2}\left[\int_0^\infty u^{\nu-\sigma}\exp\left(-\xi_1^{1/\alpha}u^{-1}\right)du\right]d\xi_1$$

$$= |z|^{\lambda+l}\Gamma(\alpha(n+l+2)-\nu)\int_{|z|}^\infty \xi_1^{-4-l+\nu/\alpha}d\xi_1$$

by (2.8) with $\delta = \sigma-\nu = (n+l+2)\alpha+1-\nu$; the equality is valid for $\nu < \alpha(n+l+2)$. The expression is finite for $\nu < (l+3)\alpha$, and is equal to

$$\frac{\Gamma(\alpha(n+l+2)-\nu)}{l+3-\nu/\alpha}|z|^{\lambda+l}|z|^{-3-l+\nu/\alpha}$$

$$= \frac{\Gamma(\alpha(n+l+2)-\nu)\Gamma(l+3-\nu/\alpha)}{\Gamma(l+4-\nu/\alpha)}|z|^{\lambda-3+\nu/\alpha}.$$

Finally, for $i \geq 2$, we have

$$\int_0^\infty |z|^\lambda u^\nu V(z,u)du$$

$$= |z|^{l+\lambda} \int_0^\infty u^{\nu-\sigma} \int_{|z|}^\infty \int_{\xi_i}^\infty \cdots \int_{\xi_2}^\infty \xi_1^{n-2} \exp(-\xi_1^{1/\alpha} u^{-1}) d\xi du$$

$$= |z|^{l+\lambda} \int_{|z|}^\infty \int_{\xi_i}^\infty \cdots \int_{\xi_2}^\infty \xi_1^{n-2} \left[\int_0^\infty u^{\nu-\sigma} \exp(-\xi_1^{1/\alpha} u^{-1}) du \right] d\xi$$

$$= |z|^{l+\lambda} \Gamma(\alpha(n+l+i+1)-\nu) \int_{|z|}^\infty \int_{\xi_i}^\infty \cdots \int_{\xi_2}^\infty \xi_1^{(-l-i-3+\nu/\alpha)} d\xi$$

taking $\delta = \alpha(n+l+i+1)-\nu+1$ in (2.8); the equality is valid for $\nu < \alpha(n+l+i+1)$. The expression is finite for $\nu < (l+3)\alpha$, and is equal to

$$\frac{|z|^{l+\lambda}\Gamma(\alpha(n+l+i+1)-\nu)|z|^{-l-3+\nu/\alpha}}{(l+3-\nu/\alpha)(l+4-\nu/\alpha)\cdots(l+i+2-\nu/\alpha)}$$

$$= \frac{\Gamma(\alpha(n+l+i+1)-\nu)\Gamma(l+3-\nu/\alpha)}{\Gamma(l+i+3-\nu/\alpha)} |z|^{\lambda-3+\nu/\alpha}.$$

Hence (2.4) holds in all cases.

For $i = 1$, note that

$$\int_0^\infty z^\lambda |u|^\nu \quad W(z,u)dz = |u|^{\nu-\sigma-2} \int_0^\infty z^{l+\lambda} \int_z^\infty \xi_1^{n-2+1/\alpha} \exp(-\xi_1^{1/\alpha}|u|^{-1}) d\xi_1 dz$$

$$= |u|^{\nu-\sigma-2} \int_0^\infty \left(\int_0^{\xi_1} z^{l+\lambda} dz \right) \xi_1^{n-2+1/\alpha} \exp(-\xi_1^{1/\alpha}|u|^{-1}) d\xi_1$$

$$= \frac{|u|^{\nu-\sigma-2}}{\lambda+l+1} \int_0^\infty \xi_1^{n+\lambda+l+1/\alpha-1} \exp(-\xi_1^{1/\alpha}|u|^{-1}) d\xi_1$$

$$= \frac{\alpha\Gamma(\alpha(n+\lambda+l)+1)}{\lambda+l+1} |u|^{\nu+(\lambda-2)\alpha-2},$$

taking $\delta = n+\lambda+l+1/\alpha-1$ in (2.7), subject to the constraint that $\lambda > -(n+\lambda+l)$.

For $i \geq 2$, we have

$$\int_0^\infty z^\lambda |u|^\nu W(z,u)dz$$

$$= |u|^{\nu-\sigma-2}\int_0^\infty z^{l+\lambda}\int_z^\infty \int_{\xi_i}^\infty \cdots \int_{\xi_3}^\infty \int_{\xi_2}^\infty \xi_1^{n-2+1/\alpha}\exp(\frac{-\xi_1^{1/\alpha}}{|u|})d\xi dz$$

$$= |u|^{\nu-\sigma-2}\int_0^\infty \xi_1^{n-2+1/\alpha}\exp(-\frac{\xi_1^{1/\alpha}}{|u|})[\int_0^{\xi_1}\cdots\int_0^{\xi_i} z^{l+\lambda}dzd\xi_i\cdots d\xi_2]d\xi_1$$

$$= |u|^{\nu-\sigma-2}\frac{\Gamma(l+\lambda+1)}{\Gamma(l+\lambda+1+i)}\int_0^\infty \xi_1^{n+\lambda+l+i-2+1/\alpha}\exp\,(-\xi_1^{1/\alpha}|u|^{-1})d\xi_1$$

$$= \frac{\alpha\Gamma(l+\lambda+l)\Gamma(\alpha(n+l+i+\lambda-1)+1)}{\Gamma(l+\lambda+1+i)}|u|^{\nu+(\lambda-2)\alpha-2}$$

taking $\delta = n+\lambda+l+i-2+1/\alpha$ in (2.7), subject to the constraint that $\delta > -1$, i.e., $\lambda > 1-(n+l+i+1/\alpha)$. Thus we have (2.5) for $i \geq 1$.

Now let $i = 1$ once again and consider

$$\int_0^\infty |z|^\lambda u^\nu W(z,u)du = |z|^{l+\lambda}\int_0^\infty u^{\nu-\sigma-2}\int_{|z|}^\infty \xi_1^{n-2+1/\alpha}\exp(-\frac{\xi_1^{1/\alpha}}{u})d\xi_1 du$$

$$= |z|^{l+\lambda}\int_{|z|}^\infty \xi_1^{n-2+\frac{1}{\alpha}}\left[\int_0^\infty u^{\nu-\sigma-2}\exp\,(-\xi_1^{1/\alpha}u^{-1})du\right]d\xi_1$$

$$= |z|^{l+\lambda}\Gamma(\alpha(n+l+2)+2-\nu)\int_{|z|}^\infty \xi_1^{-l-4+(\nu-1)/\alpha}d\xi_1$$

taking $\delta = \sigma+2-\nu = \alpha(n+l+2)+3-\nu$ in (2.8); the constraint $\delta > 1$ forces $\nu < \alpha(n+l+2)+2$. The integral is finite provided that $\nu < (l+3)\alpha+1$, and the whole expression equals

$$\frac{\Gamma(\alpha(n+l+2)+2-\nu)}{l+3-\frac{(\nu-1)}{\alpha}}|z|^{\lambda-3+(\nu-1)/\alpha}$$

$$= \frac{\Gamma(\alpha(n+l+2)+2-\nu)\Gamma(l+3+(1-\nu)/\alpha)}{\Gamma(l+4+(1-\nu)/\alpha)}|z|^{\lambda-3+(\nu-1)/\alpha}$$

when it is finite. For $i \geq 2$, we have

$$\int_0^\infty |z|^\lambda u^\nu W(z,u)du$$

$$= |z|^{l+\lambda} \int_0^\infty u^{\nu-\sigma-2} \int_{|z|}^\infty \int_{\xi_i}^\infty \cdots \int_{\xi_3}^\infty \int_{\xi_2}^\infty \xi_1^{n-2+1/\alpha} \exp(-\frac{\xi_1^{1/\alpha}}{u}) d\xi du$$

$$= |z|^{l+\lambda} \int_{|z|}^\infty \int_{\xi_i}^\infty \cdots \int_{\xi_2}^\infty \xi_1^{n-2+1/\alpha} \left[\int_0^\infty u^{\nu-\sigma-2} \exp\left(-\xi_1^{1/\alpha} u^{-1}\right) du\right] d\xi$$

$$= |z|^{l+\lambda} \Gamma(\alpha(n+l+i+1)+2-\nu) \int_{|z|}^\infty \int_{\xi_i}^\infty \cdots \int_{\xi_2}^\infty \xi_1^{-l-i-3+(\nu-1)/\alpha} d\xi$$

by (2.8) with $\delta = \sigma + 2 - \nu = \alpha(n+l+i+1) + 3 - \nu$, subject to the constraint that $\nu < \alpha(n+l+i+1)+2$. This expression is finite provided that $\nu < (l+3)\alpha + 1$, and is equal to

$$\frac{\Gamma(\alpha(n+l+i+1)+2-\nu)|z|^{l+\lambda}|z|^{-l-3+(\nu-1)/\alpha}}{[l+3+(1-\nu)/\alpha][l+4+(1-\nu)/\alpha]\cdots[l+i+2+(1-\nu)/\alpha]}$$

$$= \frac{\Gamma(\alpha(n+l+i+1)+2-\nu)\Gamma(l+3+(1-\nu)/\alpha)}{\Gamma(l+i+3+(1-\nu)/\alpha)}|z|^{\lambda-3+(\nu-1)/\alpha}.$$

This establishes (2.6) in all cases, and the proof is complete. □

Next, let $L \geq 1$ be a fixed constant, and let μ and ν be real numbers. Consider the quantities

$$\begin{aligned} V_L^\sharp(z,u) &= V_{L,\lambda,\nu,i,l}^\sharp(z,u) \\ &= \sup_{L^{-1}\leq s,t\leq L} |s^\alpha z|^\lambda |tu|^\nu V_{i,l}(s^\alpha z, tu) \end{aligned}$$

and

$$\begin{aligned} W_L^\sharp(z,u) &= W_{L,\lambda,\nu,i,l}^\sharp(z,u) \\ &= \sup_{L^{-1}\leq s,t\leq L} |s^\alpha z|^\lambda |tu|^\nu W_{i,l}(s^\alpha z, tu). \end{aligned}$$

For these 'maximal' functions, we obtain the following analogue to Lemma 2.1:

Lemma 2.2. *Let i and l be nonnegative integers. Then we have*

$$\begin{aligned} &\int_0^\infty V_L^\sharp(z,u)dz \\ (2.9)\qquad &\leq L^{6(\sigma+|\lambda|+|\nu|)} \frac{\alpha\Gamma(\alpha(n+l+\lambda+i-1))\Gamma(l+\lambda+1)}{\Gamma(l+\lambda+i+1)} |u|^{\nu+(\lambda-2)\alpha-1} \end{aligned}$$

for all $\lambda > 1-(n+l+i)$ and for all real ν; and

$$\begin{aligned} &\int_0^\infty V_L^\sharp(z,u)du \\ (2.10)\qquad &\leq L^{6(\sigma+|\lambda|+|\nu|)} \frac{\Gamma(\alpha(n+l+i+1)-\nu)\Gamma(l+3-\nu/\alpha)}{\Gamma(l+i+3-\nu/\alpha)} |z|^{\lambda-3+\nu/\alpha} \end{aligned}$$

for all $\lambda \in R$ *and for all* $\nu < (l+3)\alpha$. *Moreover, for* $i \geq 1$ *we also have*

$$\begin{aligned}(2.11)\qquad &\int_0^\infty W_L^\sharp(z,u)dz \\ &\leq L^{6(\sigma+|\lambda|+|\nu|)} \frac{\alpha\Gamma(l+\lambda+1)\Gamma(\alpha(n+l+i+\lambda-1)+1)}{\Gamma(l+\lambda+i+1)} |u|^{\nu+(\lambda-2)\alpha-2}\end{aligned}$$

for all $\lambda > 1-(n+l+i+1/\alpha)$ *and for all real* ν; *and*

$$\begin{aligned}(2.12)\qquad &\int_0^\infty W_L^\sharp(z,u)du \\ &\leq L^{6(\sigma+|\lambda|+|\nu|)} \frac{\Gamma(\alpha(n+l+i+1)+2-\nu)\Gamma(l+3+\frac{1-\nu}{\alpha})}{\Gamma(l+i+3+\frac{1-\nu}{\alpha})} |z|^{\lambda-3+\frac{\nu-1}{\alpha}}\end{aligned}$$

for all $\lambda \in R$ *and for all* $\nu < (l+3)\alpha + 1$.

Proof. The result is essentially a corollary of Lemma 2.1 and uses two elementary facts: first, for p real, $L^{-1} < s < L$, we have $s^p \leq L^{|p|}$; second, for $p > 0$, $L^{-1} < s < L$, we have $\exp(-ps) \leq \exp(-pL^{-1})$. From this it is easily seen that

$$(2.13)\qquad V_L^\sharp(z,u) \leq L^\gamma |\zeta|^\lambda |\mu|^\nu V(\zeta,\mu),\ \ W_L^\sharp(z,u) \leq L^\delta |\zeta|^\lambda |\mu|^\nu W(\zeta,\mu),$$

where $\zeta = L^{-\alpha}z$, $\mu = Lu$, $\delta = 2\alpha|\lambda+l| + 2|\nu-\sigma-2|$, and

$$\gamma = \begin{cases} 2\alpha|\lambda+n+l-2| + 2|\nu-\sigma|, & i = 0, \\ 2\alpha|\lambda+l| + 2|\nu-\sigma|, & i \geq 1. \end{cases}$$

An extremely crude estimate shows that, for each fixed choice of i and l, both γ and δ are less than or equal to $6\sigma + 2(|\lambda|+|\nu|)$. The inequalities (2.9)-(2.12) now follow from (2.3)-(2.6) upon integration of (2.13). □

Next, we introduce yet another family of functions, built up from V, and obtain estimates for them which are crucial to what follows. Let i, l be fixed nonnegative integers and let L, M, N be positive constants with $L \geq 1$. Define

(2.14a)

$$\mathcal{H}_1(z,u) = \sup_{L^{-1}\leq s,t\leq L} \chi_{[-M,M]}(tu)[1-\chi_{(-N^\alpha,N^\alpha)}(s^\alpha z)] \int_0^M \tau^{2\alpha} V(s^\alpha z,\tau)d\tau,$$

(2.14b)

$$\mathcal{H}_2(z,u) = \sup_{L^{-1}\leq s,t\leq L} \chi_{[-M^\alpha,M^\alpha]}(s^\alpha z)[1-\chi_{(-N,N)}(tu)]|tu|^{2\alpha} \int_0^{M^\alpha} V(w,tu)dw.$$

Then we have

Lemma 2.3. *With $\mathcal{H}_1, \mathcal{H}_2$ as above,*

(2.15a)
$$\int_{N^\alpha}^{\infty} \mathcal{H}_1(z,u)(1+z^{1/2})dz$$

$$\leq 4L^{6(\sigma+3)}M^{\alpha/2}\max\{1, N^{-\alpha/2}\}\frac{\Gamma(\alpha(n+l+i-1/2))\Gamma(l+3/2)}{\Gamma(l+i+3/2)},$$

when $u \in [-M, M]$ and

(2.15b)
$$\int_{N}^{\infty} \mathcal{H}_2(z,u)(1+u^{\alpha/2})du$$

$$\leq 4L^{6(\sigma+3)}M^{\alpha/2}\max\{1, N^{-a/2}\}\frac{\Gamma(\alpha(n+l+i-3/2))\Gamma(l+1/2)}{\Gamma(l+i+1/2)},$$

when $z \in [-M^\alpha, M^\alpha]$.

Proof. First, note that if $z \geq N^\alpha$ then $1+z^{1/2} \leq 2\max\{1, N^{-\alpha/2}\}z^{1/2}$, while if $u \geq N$, then $1+u^{\frac{\alpha}{2}} \leq 2\max\{1, N^{-\frac{\alpha}{2}}\}u^{\alpha/2}$. Taking $\lambda = 1/2$ and $\nu = 2\alpha$ in (2.9) , we have, for $u \in [-M, M]$,

$$\int_{N^\alpha}^{\infty} \mathcal{H}_1(z,u)z^{\frac{1}{2}}dz \leq \int_{N^\alpha}^{\infty}\int_0^M V_L^\sharp(z,\tau)d\tau dz \leq \int_0^M\int_0^\infty V_L^\sharp(z,\tau)dzd\tau$$

$$\leq L^{6(\sigma+3)}\frac{\alpha\Gamma(\alpha(n+l+i-1/2))\Gamma(l+3/2)}{\Gamma(l+i+3/2)}\int_0^M \tau^{\frac{\alpha}{2}-1}d\tau$$

from which (2.15a) follows upon integration. For $z \in [-M^\alpha, M^\alpha]$, we have

$$\int_N^\infty \mathcal{H}_2(z,u)u^{\frac{\alpha}{2}}du \leq \int_N^\infty\int_0^{M^\alpha} V_L^\sharp(w,u)dwdu$$

$$\leq \int_0^{M^\alpha}\int_0^\infty V_L^\sharp(w,u)dudw$$

$$\leq L^{6(\sigma+3)}\frac{\Gamma(\alpha(n+l+i-\frac{3}{2}))\Gamma(l+\frac{1}{2})}{\Gamma(l+i+\frac{1}{2})}\int_0^M \tau^{\frac{\alpha}{2}-1}d\tau,$$

by (2.10) with $\lambda = 0$, $\nu = 5\alpha/2$; (2.15a) follows upon integration. □

Next, we obtain estimates on the partial derivatives of V. By straightforward calculation, it is easy to see that

$$\left|\tfrac{\partial}{\partial z}V_{i,l}(z,u)\right| \leq K_1(z,u) = K_{1,i,l}(z,u)$$

(2.16)
$$= \begin{cases} [\,|z|^{-1}(l+n-2)+\alpha^{-1}|z|^{1/\alpha-1}|u|^{-1}]V_{i,l}(z,u), & i=0,\\ l|z|^{-1}V_{i,l}(z,u)+|u|^{-\alpha}V_{i-1,l}(z,u), & i\geq 1.\end{cases}$$

$$\left|\frac{\partial}{\partial u}V_{i,l}(z,u)\right| \le K_2(z,u) = K_{2,i,l}(z,u)$$

(2.17)
$$= \begin{cases} [\sigma(i,l)|u|^{-1} + |u|^{-2}|z|^{1/\alpha}]V_{i,l}(z,u), & i=0, \\ \sigma(i,l)|u|^{-1}V_{i,l}(z,u) + W_{i,l}(z,u), & i \ge 1. \end{cases}$$

We also note, at this juncture, that for $\lambda \ge 0$ and for $z,\ u \in R$,

$$V(\lambda^\alpha z, \lambda u) = \lambda^{-1-3\alpha} V(z,u), \tag{2.18a}$$

$$K_1(\lambda^\alpha z, \lambda u) = \lambda^{-1-4\alpha} K_1(z,u), \tag{2.18b}$$

$$K_2(\lambda^\alpha z, \lambda u) = \lambda^{-2-3\alpha} K_2(z,u) \tag{2.18c}$$

i.e., V, K_1, and K_2 are homogeneous of degree $-1-3\alpha$, $-1-4\alpha$, and $-2-3\alpha$, respectively, relative to the nonisotropic dilations (1.1).

Finally, we obtain estimates for an important family of function built up from V, K_1, and K_2. We define, for $c > 0$ and for nonnegative integers i and l, the quantity

$$\beta_1 = \beta_1(i,l,n,\alpha,c) = c^{2i+3l+1}\frac{\Gamma(\alpha(i+l+n)+4)\Gamma(l+4)}{\Gamma(i+l+1)}$$

If we wish to indicate the dependence of β_1 upon one or more of the parameters, we indicate those parameters while suppressing the others. Most often, we shall emphasize the dependence upon the positive constant c by writing $\beta_1(c)$.

Let L, M, N be positive constants, with $L \ge 1$, and let i and l be nonnegative integers, as before. Define

(2.19)
$$\begin{aligned} \mathcal{H}_3(z,u) &= \sup_{L^{-1}\le s,t\le L} [1-\chi_{(-M^\alpha,M^\alpha)}(s^\alpha z)][1-\chi_{(-N,N)}(tu)] \\ &\cdot\{|tu|^\alpha V(s^\alpha z, tu) + |tu|^{2\alpha}(K_1(s^\alpha z,tu) + K_2(s^\alpha z,tu))\}. \end{aligned}$$

Then we have

Proposition 2.4. *There is a constant $c > 0$ depending on α, n, L, M, N such that, for $j = 1,2,3$,*

$$\iint_{R^2} \mathcal{H}_j(z,u)(1+|u|^{\alpha/2}+|z|^{1/2})\,du\,dz \le \beta_1(c). \tag{2.20}$$

Proof. For $j = 1,2$, (2.20) is an immediate consequence of Lemma 2.3 and the definition (2.14) of the functions $\mathcal{H}_j$. For $j = 3$, (2.20) follows from (2.19) and (2.16)-(2.17) by repeated application of Lemma 2.2, especially (2.9) and (2.11). The proof is very much like that of Lemma 2.3 and is left to the reader. □

We note that the 'omnibus estimate' (2.20), while not as sharp as our previous estimates, is more than adequate for our purposes. With this estimate in hand, we are ready for our first application of the T1 Theorem. For the balance of this section, let q be a real-valued compactly-supported function on R^2 satisfying (0.1)-(0.4) with f replaced by q and n replaced by 2.

Proposition 2.5. *For $g \in C_0^\infty(R^2)$ and for fixed $i, l, n, a_1, a_2, \alpha$ as above, set*

$$Sg(x,t) = \int_R \int_R \Psi(x,t,y,s) g(y,s)\,dyds, \quad (x,t) \in R^2, \tag{2.21}$$

where

$$\Psi(x,t,y,s) = [(q(x,s) - q(x,t))^2 + (q(y,s) - q(y,t))^2] V(x-y, s-t) \tag{2.22}$$

for $x, t, y, s \in R$. Then S extends to a bounded operator on $L^2(R^2)$, and there is a constant $c_2 > 0$ (depending upon α and n) such that

$$||Sg||_2 \leq \beta_1(c_2)(a_1^2 + a_2^2)||g||_2. \tag{2.23a}$$

Moreover, if $g \in L^2(R^2) \cap L^\infty(R^2)$ or $g \equiv 1$ then

$$||Sg||_{*,\alpha} \leq \beta_1(c_2)(a_1^2 + a_2^2)||g||_\infty \tag{2.23b}$$

where S1 is defined by taking $g \equiv 1$ in (2.21).

Proof. To begin, observe that Ψ is nonnegative and S is self-adjoint. Next, we show that S1 ($= S^*1$) is in BMO$_\alpha$ (from which it follows that (1.4) holds, with Ψ in place of H, for almost every $(x,t) \in R^2$). To do this, we make use of the results of this section, together with the Lipschitz estimates (0.1) and (0.4) for q, together with the following fact due to Strichartz ([Str]): for every $x \in R$, the measure

$$d\mu_x(s,t) = \frac{|q(x, s+t) - q(x,s)|^2}{|t|^{1+2\alpha}} dsdt \tag{2.24a}$$

is a Carleson measure on R_+^2, i.e., there is a constant $c > 0$ such that

$$\mu_x(\{(s,t) : \ |s - s_0| < d, \ |t| < d\}) \leq c a_2^2 d, \ s_0 \in R, \ d > 0. \tag{2.24b}$$

Given $x_0, t_0 \in R$ and $d > 0$, let $Q = Q_d(x_0, t_0)$ be as in (1.2a). Let $\tilde{Q} = Q_{Kd}(x_0, t_0)$ where $K = 2 \cdot 6^{1/\alpha^2}$. For ease of notation we write $\tilde{Q} = I_1 \times I_2$, where

$$I_1 = \{x : \ |x - x_0| < 6^{1/\alpha} d^\alpha\}, \ I_2 = \{t : \ |t - t_0| < 6^{1/\alpha^2} d\}.$$

For $(x,t) \in Q$, we write

$$\begin{aligned} S1(x,t) &= \iint_{\tilde{Q}} \Psi(x,t,y,s)\,dyds + \iint_{R^2 \setminus \tilde{Q}} \Psi(x,t,y,s)\,dyds \\ &= A_1(x,t) + A_2(x,t). \end{aligned} \tag{2.25}$$

Then, using (2.24) together with (2.3) for $\lambda = \nu = 0$, we have

$$
\begin{aligned}
&\iint\limits_{Q} |A_1(x,t)|dxdt \\
&= \iint\limits_{Q} \{ \iint\limits_{Q} [(q(x,s)-q(x,t))^2 + (q(y,s)-q(y,t))^2] \\
&\qquad\qquad \cdot V(x-y,s-t)dyds\}dxdt \\
&\le \iint\limits_{Q} \{ \int_{-Kd}^{Kd} \int_{I_1} [(q(x,u+t)-q(x,t))^2 + (q(y,u+t)-q(y,t))^2] \\
&\qquad\qquad \cdot V(x-y,u)dydu\}dxdt \\
&\le 2 \iint\limits_{Q} \left\{ \int_{-Kd}^{Kd} (q(x,u+t)-q(x,t))^2 \int_{R} V(x-y,u)dydu \right\} dxdt \\
&\le \frac{4\alpha\Gamma(\alpha(n+l+i-1))\Gamma(l+1)}{\Gamma(l+i+1)} \int_{I_1}\int_{-Kd}^{Kd}\int_{I_2} d\mu_x(t,u)dx \\
&\le c_\alpha a_2^2 \frac{\Gamma(\alpha(n+l+i-1))\Gamma(l+1)}{\Gamma(l+i+1)} d^{1+\alpha} \le \beta_1(c_{\alpha,n}) a_2^2\, d^{1+\alpha}
\end{aligned}
\tag{2.26}
$$

for c sufficiently large. Next from (2.3) of Lemma 2.1 and the fact that q has compact support we deduce, $A_2(x,t) < \infty$, for each $(x,t) \in R^2$. From this deduction and (2.26) we see that (1.4) is valid with H replaced by Ψ.

To obtain better estimates on A_2, we write $R^2 \setminus \tilde{Q} = E_1 \cup E_2 \cup E_3$, where

$$E_1 = \left\{(y,s) : |y-x_0| \ge \left(\tfrac{K}{2}d\right)^\alpha, |s-t_0| \le \tfrac{K}{2}d\right\} = (R\setminus I_1)\times I_2$$

$$E_2 = \left\{(y,s) : |y-x_0| < \left(\tfrac{K}{2}d\right)^\alpha, |s-t_0| \ge \tfrac{K}{2}d\right\} = I_1\times (R\setminus I_2)$$

$$E_3 = \left\{(y,s) : |y-x_0| \ge \left(\tfrac{K}{2}d\right)^\alpha, |s-t_0| \ge \tfrac{K}{2}d\right\} = (R\setminus I_1)\times (R\setminus I_2)$$

with $K = 2\cdot 6^{1/\alpha^2}$, as before.

Now note that, by (0.4), when $(x,t) \in Q$,

$$\begin{aligned}\iint\limits_{E_1} \Psi(x,t,y,s)dyds &\leq ca_2^2 \iint\limits_{E_1} |s-t|^{2\alpha} V(x-y,s-t)dyds \\
&\leq ca_2^2 \int_{(Pd)^\alpha}^{\infty} \int_0^{K+1)d/2} u^{2\alpha} V(z,u)dudz \qquad (2.27)\\
&= ca_2^2 \int_{P^\alpha}^{\infty} \int_0^{(K+1)/2} \tau^{2\alpha} V(\zeta,\tau)d\zeta d\tau\end{aligned}$$

where $P^\alpha = \left(\frac{K}{2}\right)^\alpha - 2^{-\alpha}$ and we make the change of variables $\tau = ud^{-1}$, $\zeta = zd^{-\alpha}$ and use (2.18a). By (2.14a) and (2.15a) with $L = 1$, $M = (K+1)/2$, and $N = P$,

$$(2.28) \quad \iint\limits_{E_1} \Psi(x,t,y,s)\, dyds \leq ca_2^2 \int_{P^\alpha}^{\infty} \mathcal{H}_1(\zeta,0)d\zeta \leq \beta_1(c_{\alpha,n})a_2^2, \ (x,t) \in Q$$

for $c_{\alpha,n}$ large enough. Similarly, by (0.4), when $(x,t) \in Q$ we have

$$\begin{aligned}\iint\limits_{E_2} \Psi(x,t,y,s)dyds &\leq ca_2^2 \int_{(K-1)d/2}^{\infty} \int_0^{(Pd)^\alpha} u^{2\alpha} V(z,u)dzdu \\
&\leq ca_2^2 \int_{(K-1)/2}^{\infty} \mathcal{H}_2(0,\tau)d\tau \qquad (2.29)\\
&\leq \beta_1(c_{\alpha,n})a_2^2, \ (x,t) \in Q,\end{aligned}$$

where this time $P^\alpha = \left(\frac{K}{2}\right)^\alpha + 2^{-\alpha}$, and we use (2.18a), (2.14b), and (2.15b) with $L = 1$, $M = P$, and $N = (K-1)/2$.

Finally, suppose that $(x,t) \in Q$ and $(y,s) \in E_3$, and note that

$$\begin{aligned}&|\Psi(x,t,y,s) - \Psi(x_0,t_0,y,s)| \\
&\leq |(q(x,s)-q(x,t))^2 + (q(y,s)-q(y,t))^2 - (q(x_0,s)-q(x_0,t_0))^2 \\
&+ (q(y,s)-q(y,t_0))^2|V(x-y,s-t) + [(q(x_0,s)-q(x_0,t_0))^2 \\
&+ (q(y,s)-q(y,t_0))^2]|\nabla V(x'-y,s-t')\cdot(x_0-x,t_0-t)| \\
&\leq c(a_1^2+a_2^2)d^\alpha(|s-t|^\alpha + |s-t_0|^\alpha)V(x-y,s-t) \\
&\qquad + ca_2^2|s-t_0|^{2\alpha}(d^\alpha K_1(x'-y,s-t') + dK_2(x'-y,s-t')),\end{aligned}$$

for some x' between x and x_0, and some t' between t and t_0. Here we have used (0.1), (0.4), (2.16)-(2.18), and the mean-value theorem of calculus. It is easily verified that

$$1 + K^{-\alpha} \geq \frac{|x'-y|}{|x_0-y|} \geq 1 - K^{-\alpha}, \ 1 + K^{-1} \geq \frac{|s-t'|}{|s-t_0|} \geq 1 - K^{-1},$$

for (x,t), $(x',t') \in Q$ and $(y,s) \in E_3$. Moreover, it is easily verified that $(1-K^{-\alpha})^{-1} \geq 1+K^{-\alpha}$ and $(1-K^{-1})^{-1} \geq 1+K^{-1}$, so if we take $L = \max\{(1-K^{-1})^{-1},\ (1-K^{-\alpha})^{-1/\alpha}\}$, then we have

$$L^\alpha|x_0-y| \geq |x'-y| \geq L^{-\alpha}|x_0-y|,\ L|s-t_0| \geq |s-t'| \geq L^{-1}|s-t_0|,$$

for $(x,t),(x',t') \in Q$ and $(y,s) \in E_3$. For this choice of L and for $M^\alpha = (K^\alpha-1)2^{-\alpha}$, $N = (K-1)/2$, we have, for $(x,t) \in Q$ and $(y,s) \in E_3$,

$$|\Psi(x,t,y,s) - \Psi(x_0,t_0,y,s)| \leq c(a_1^2+a_2^2)(\mathcal{H}_3)_d(x_0-y,s-t_0).$$

By (2.20) with $j=3$ we obtain

$$\iint\limits_{E_3} |\Psi(x,t,y,s) - \Psi(x_0,t_0,y,s)|dyds$$

(2.30)
$$\leq c(a_1^2+a_2^2)\int_{Nd}^{\infty}\int_{(Md)^\alpha}^{\infty} (\mathcal{H}_3)_d(x_0-y,s-t_0)dyds$$

$$\leq \beta_1(c_{\alpha,n})(a_1^2+a_2^2),\ (x,t) \in Q.$$

Combining (2.28)-(2.30), we have

(2.31)
$$|A_2(x,t) - A_2(x_0,t_0)| \leq \beta_1(c)(a_1^2+a_2^2),\ (x,t) \in Q.$$

By (2.25), (2.26), and (2.31), we have

$$\iint\limits_{Q} |(S1)(x,t) - m_Q(S1)|dxdt$$

(2.32)
$$\leq 2\iint\limits_{Q} |A_1(x,t)|dxdt + 2\iint\limits_{Q} |A_2(x,t) - A_2(x_0,t_0)|dxdt$$

$$\leq \beta_1(c)(a_1^2+a_2^2)|Q|,$$

for c sufficiently large. Since Q is arbitrary, it follows that $S1 = S^*1$, is in BMO_α and

(2.33)
$$||S1||_{*,\alpha} \leq \beta_1(c)(a_1^2+a_2^2).$$

Thus (1.11) of the T1 Theorem is satisfied with $a_5 = \beta_1(c)(a_1^2+a_2^2)$. In particular, we obtain (1.4), with Ψ in place of H, for almost every $(x,t) \in R^2$. Moreover, (2.33) together with the compact support of q imply that S maps $C_0^\infty(R^2)$ boundedly into $L^2(R^2)$.

Since S is self-adjoint, we need only verify (1.7)-(1.9) for S in place of T and an appropriate choice of h. Let $\lambda = d/2$ and let θ be defined as in section 1. We note that the support of $\theta_\lambda^{(x_0,t_0)}$ is contained in Q. Now let $P^\alpha = (K^\alpha+1)2^{-\alpha}$; note that $P \geq (K+1)/2$. Let

$$Q^\sharp = Q_{2Pd}(x_0,t_0) = \{(x,t) :\ |x-x_0| < (Pd)^\alpha,\ |t-t_0| < Pd\}.$$

Now, if $|x_0 - y_0| < (K/2)^\alpha d^\alpha$ and $|t_0 - s_0| < Kd/2$, then the support of $\theta_\lambda^{(y_0,s_0)}$ is contained in $Q^\sharp$, and

(2.34)
$$
\begin{aligned}
&|\langle (D\theta\ )_\lambda^{(x_0,t_0)}, S(\theta_\lambda^{(y_0,s_0)})\rangle| \\
&\leq \iint_Q |(D\theta)_\lambda^{(x_0,t_0)}(x,t)| \iint_{Q^\sharp} \Psi(x,t,y,s)\theta_\lambda^{(y_0,s_0)}(y,s)dydsdxdt \\
&\leq c\lambda^{-2-2\alpha} \iint_{Q^\sharp} |A_1(z,\tau)|dzd\tau \leq \beta_1(c)a_2^2\lambda^{-1-\alpha}
\end{aligned}
$$

as in (2.26), for c large enough. Otherwise, for some $j \in \{1,2,3\}$ we have

$$
\operatorname{supp}(\theta_\lambda^{(y_0,s_0)}) \subseteq E_j. \tag{2.35}
$$

If (2.35) holds for $j = 1$ or 2, we have

(2.36)
$$
\begin{aligned}
&|\langle (D\theta\ )_\lambda^{(x_0,t_0)}, S(\theta_\lambda^{(y_0,s_0)})\rangle| \\
&\leq ca_2^2\lambda^{-1-\alpha} \sup_{(x,t)\in Q} \iint_{E_j \cap Q_d(y_0,s_0)} |s-t|^{2\alpha} V(x-y, s-t)dyds.
\end{aligned}
$$

For $j = 1$, note that if $(x,t) \in Q$ and $(y,s) \in E_1 \cap Q_d(y_0,s_0)$, then $|s-t| \leq M_1 d$ and $|x-y| \geq N_1^\alpha d^\alpha$, where $M_1 = (K+1)/2$ and $N_1^\alpha = (K^\alpha - 1)2^{-\alpha}$; moreover

$$
0 < 1 - 2(K^\alpha - 1)^{-1} \leq \frac{|x-y|}{|x_0 - y_0|} \leq 1 + 2(K^\alpha - 1)^{-1}.
$$

As before, it is easily checked that $[1 - 2(K^\alpha - 1)^{-1}]^{-1} > 1 + 2(K^\alpha - 1)^{-1}$, so if we take $L_1 = [1 - 2(K^\alpha - 1)^{-1}]^{-1/\alpha}$, then we have

(2.37)
$$
\begin{aligned}
&|\langle (D\theta)_\lambda^{(x_0,t_0)}, S(\theta_\lambda^{(y_0,s_0)})\rangle| \\
&\leq ca_2^2\lambda^{-1-\alpha} \sup_{L_1^{-1}\leq\mu\leq L_1} \iint_{E_1 \cap Q_d(y_0,s_0)} |s-t|^{2\alpha} V(\mu^\alpha(x_0 - y_0), s-t)dyds \\
&\leq ca_2^2\lambda^{-1} \sup_{L_1^{-1}\leq\mu\leq L_1} \int_{\{|s-t|\leq M_1 d\}} |s-t|^{2\alpha} V(\mu^\alpha(x_0 - y_0), s-t)ds \\
&\leq ca_2^2(\mathcal{H}_1)_\lambda(x_0 - y_0, s_0 - t_0).
\end{aligned}
$$

where we have used (2.18a) and $\mathcal{H}_1$ is defined by (2.14a) with $L = L_1$, $M = M_1$, $N = N_1$.

For $j = 2$, note that if $(x,t) \in Q$ and $(y,s) \in E_2 \cap Q_d(y_0, s_0)$, then $|x-y| \le M_2^\alpha d^\alpha$ and $|s-t| \ge N_2 d$, where $M_2^\alpha = (K^\alpha + 1)2^{-\alpha}$ and $N_2 = (K-1)/2$; moreover,

$$0 < 1 - 2(K-1)^{-1} \le \frac{|s-t|}{|s_0 - t_0|} \le 1 + 2(K-1)^{-1};$$

as before $[1-2(K-1)^{-1}]^{-1} \ge 1+2(K-1)^{-1}$, so if we take $L_2 = [1-2(K-1)^{-1}]^{-1}$, we have

(2.38)

$$\begin{aligned} &|\langle (D\theta)_\lambda^{(x_0,t_0)}, S(\theta_\lambda^{(y_0,s_0)}\rangle| \\ &\le c a_2^2 \lambda^{-1-\alpha} \sup_{L_2^{-1} \le \mu \le L_2} \iint_{E_2 \cap Q_d(y_0,s_0)} |\mu(s_0 - t_0)|^{2\alpha} V(x-y, \mu(s_0-t_0)) dyds \\ &\le c a_2^2 \lambda^{-\alpha} \sup_{L_2^{-1} \le \mu \le L_2} |\mu(s_0-t_0)|^{2\alpha} \int_{\{|x-y| \le M_2^\alpha d^\alpha\}} V(x-y, \mu(s_0-t_0)) dy \\ &\le c a_2^2 (\mathcal{H}_2)_\lambda (x_0 - y_0, s_0 - t_0)\,, \end{aligned}$$

where we have used (2.18a) and $\mathcal{H}_2$ is defined by (2.14b) with $L = L_2$, $M = M_2$, $N = N_2$.

If (2.35) holds for $j = 3$, we make use of the fact that

$$\iint_{R^2} (D\theta)_\lambda^{(x_0,t_0)}(y,s) dyds = 0, \tag{2.39a}$$

$$\iint_{R^2} \theta_\lambda^{(y_0,s_0)}(y,s) dyds = 1, \tag{2.39b}$$

from which it follows that

(2.40)

$$\begin{aligned} &|\langle (D\theta)_\lambda^{(x_0,t_0)}, S(\theta_\lambda^{(y_0,s_0)})\rangle| \\ &= | \iint_Q (D\theta_\lambda)^{(x_0,t_0)}(x,t) \iint_{E_3 \cap Q_d(y_0,s_0)} [\Psi(x,t,y,s) - \Psi(x_0,t_0,y_0,s_0)] \\ &\qquad \cdot \theta_\lambda^{(y_0,s_0)}(y,s) dydsdxdt| \\ &\le c\lambda^{-1-\alpha} \sup_{(x,t)\in Q} \iint_{E_3 \cap Q_d(y_0,s_0)} |\Psi(x,t,y,s) - \Psi(x_0,t_0,y_0,s_0)| dyds. \end{aligned}$$

We argue here as in (2.30): for $(x,t) \in Q$ and $(y,s) \in E_3 \cap Q_d(y_0,s_0)$, we have

$$\begin{aligned} &|\Psi(x,t,y,s) - \Psi(x_0,t_0,y_0,s_0)| \\ &\le |\Psi(x,t,y,s) - \Psi(x_0,t_0,y,s)| + |\Psi(x_0,t_0,y,s) - \Psi(x_0,t_0,y_0,s_0)| \\ &\le c(a_1^2 + a_2^2)(\mathcal{H}_3)_\lambda(x_0 - y_0, s_0 - t_0), \end{aligned}$$

where $\mathcal{H}_3$ is given by (2.19) with $M^\alpha = M_3^\alpha = (K^\alpha - 1)2^{-\alpha}$, $N = N_3 = (K-1)/2$, and now

$$L = L_3 = \max\{[1 - 2(K^\alpha - 1)^{-1}]^{-1/\alpha},\ [1 - 2(K-1)^{-1}]^{-1}\}.$$

Then, by (2.40), we easily obtain

$$\begin{aligned} &|\langle (D\theta)_\lambda^{(x_0,t_0)},\ S(\theta_\lambda^{(y_0,s_0)})\rangle| \\ &\qquad \le c(a_1^2 + a_2^2)(\mathcal{H}_3)_\lambda(x_0 - y_0, s_0 - t_0). \end{aligned} \tag{2.41}$$

Now let $\mathcal{H}_0$ denote the characteristic function of the set

$$\{(z,\tau) :\ |z| \le K^\alpha,\ |\tau| \le K\} = Q_{2K}(0,0)\,.$$

If we set

$$\mathcal{H} = \beta_1(c)a_2^2\mathcal{H}_0 + ca_2^2\mathcal{H}_1 + ca_2^2\mathcal{H}_2 + c(a_1^2 + a_2^2)\mathcal{H}_3$$

where c is the largest of the constants appearing in (2.34), (2.37), (2.38), and (2.41), then we have

$$|\langle (D\theta)_\lambda^{(x_0,t_0)},\ S(\theta_\lambda^{(y_0,s_0)})\rangle| \le \mathcal{H}_\lambda(x_0 - y_0, s_0 - t_0). \tag{2.42}$$

Now, if we set

$$a_3 = \iint\limits_{R^2} (|y| + |s|^\alpha + 1)^{1/2}\mathcal{H}(y,s)dyds,$$

then, by (2.20), we have $a_3 \le \beta_1(c)(a_1^2 + a_2^2)$. Taking $h = a_3^{-1}\mathcal{H}$, we obtain (1.7)-(1.10) of the T1 Theorem with S in place of T, $\delta = 1/2$, a_3 as above, and $a_4 = a_5$. The L^2 estimate (2.23a) now follows from (1.12) of the T1 Theorem.

It remain to establish (2.23b) for $g \in L^2(R^2) \cap L^\infty(R^2)$. We sketch the proof, which follows a well-known argument due to Peetre. Write $g = g_1 + g_2$, where $g_1 = g\chi_{\tilde{Q}}$. Then, by (2.23a) and the Schwartz inequality,

$$\begin{aligned} \iint\limits_{Q} |Sg_1(x,t)|dxdt &\le (2^{1-d}d^{1+\alpha})^{1/2}\,||Sg||_2 \\ &\le \beta_1(c)(a_1^2 + a_2^2)(d^{\alpha+1})^{1/2}\,||g_1||_2 \\ &\le |Q|\,\beta_1(c)\,(a_1^2 + a_2^2)\,||g||_\infty. \end{aligned}$$

Moreover, as in (2.30) we deduce

$$|Sg_2(x,t) - Sg_2(x_0,t_0)| \le \beta_1(c)\,(a_1^2 + a_2^2)\,||g||_\infty,\ (x,t) \in Q.$$

Combining these two estimates as in (2.32) and (2.33), we obtain (2.23b), and the proof is complete. □

3. Estimates for a Recursively-Defined Family of Operators.

In this section we continue our progression toward Theorem 1 by obtaining estimates for a still more complicated family of operators than that considered in Proposition 2.5. We continue to use the notation of the preceding sections, and we introduce some additional notation, as follows.

As in section 1, ϕ denotes the function given by

$$\phi(u) = \chi_{(-1,1)}(u)\exp\left[-(1-u^2)^{-1}\right],\ u \in R.$$

For $\lambda > 0$ and $z \in R$, we define

$$\phi_\lambda^z(x) = \|\phi\|_1^{-1}\lambda^{-1}\phi\left(\frac{z-x}{\lambda}\right),\quad x \in R;$$

ϕ_λ^z is the (one-dimensional) translate and dilate of ϕ, normalized to have L^1 norm equal to 1. Let γ be a nonnegative C^∞ function supported in $(1/2,\infty)$ and identically one on $[1,\infty)$, having the property that $|\gamma'| \le 1000$.

Unless noted otherwise, all sequences in this section are assumed to be indexed by the natural numbers N. For the balance of this section we consider three fixed sequences: a sequence $\langle z_j\rangle$ of real numbers; a sequence $\langle\lambda_j\rangle$ of positive numbers; and a sequence $\langle\mu_j\rangle$ such that, for each $j \in N, \mu_j$ is equal to either 1 or λ_j^α. From these three sequences we construct three additional sequences: $\langle\psi_j\rangle$ is a sequence of functions given by

$$\psi_j = \begin{cases} 1, & \text{if } \mu_j = 1, \\ \phi_{\mu_j}^{z_j}, & \text{if } \mu_j = \lambda_j^\alpha; \end{cases}$$

$\langle\delta_j\rangle$ is a sequence of numbers indexed by nonnegative integers, given by

$$\delta_0 = 1,\ \delta_j = \mu_1\mu_2\cdots\mu_j \text{ if } j \ge 1;$$

$\eta = \langle\eta_j\rangle$ is a sequence of pairs such that, for each $j \in N$, either $\eta_j = (\psi_j,+)$ or $\eta_j = (\psi_j,-)$. Finally, when i and j are natural numbers with $1 \le j \le i$, we write

$$\hat{\delta}_{i,j} = \delta_i/\mu_j = \mu_1\mu_2\cdots\hat{\mu}_j\cdots\mu_i,$$

where the carat in the rightmost expression denotes the deletion of μ_j from the product. These definitions are intentionally imprecise, so as to cover a range of possibilities.

As before, let n be an integer greater than 1; let $0 < \epsilon < N$. For $j \in N \cup \{0\}$ we define $\Omega_j = \Omega_{j,\eta,n,\epsilon,N} :\ R^3 \to R$ recursively by setting

$$\Omega_0(x,y,u) = \gamma(|x-y|\epsilon^{-1})[1-\gamma(|x-y|N^{-1})]|x-y|^{n-2}\exp\left[-|x-y|^{1/\alpha}|u|^{-1}\right];$$

for $j \in N$, if $\eta_j = (\psi_j, +)$, we let

$$\Omega_j(x,y,u) = \begin{cases} \displaystyle\int_y^\infty \psi_j(\xi)\Omega_{j-1}(x,\xi,u)d\xi, & x < y,\ u \in R, \\ \displaystyle -\int_{-\infty}^y \psi_j(\xi)\Omega_{j-1}(x,\xi,u)d\xi, & y < x, u \in R; \end{cases}$$

while if $\eta_j = (\psi_j, -)$, then we let

$$\Omega_j(x,y,u) = \begin{cases} \displaystyle\int_y^\infty \psi_j(\xi)\Omega_{j-1}(\xi,x,u)d\xi, & x < y,\ u \in R, \\ \displaystyle -\int_{-\infty}^y \psi_j(\xi)\Omega_{j-1}(\xi,x,u)d\xi, & y < x,\ u \in R. \end{cases}$$

For each pair i, l of nonnegative integers, and for each measurable function $p:\ R^4 \to R$, we let $[p]$ stand for the operator given by

$$[p]g(x,t) = \iint\limits_{R^2} |s-t|^{-[(n+l+i-1)\alpha+1]} p(x,t,y,s)\Omega_i(x,y,s-t)g(y,s)dyds$$

when $g \in C_0^\infty(R^2)$ and $(x,t) \in R^2$. $[p]$ depends upon a whole host of parameters (i, l, n, η, N); when we wish to indicate the dependence of $[p]$ upon one or more specific parameters, we write these parameters as subscripts, suppressing the others. We say that the function p is the representative for the operator $[p]$, and we identify functions from which the representatives of concern to us are built:

$$p_1(x,t,y,s) = q(y,s) - q(x,s),$$

$$p_2(x,t,y,s) = q(y,t) - q(x,t) = p_1(x,s,y,t),$$

$$p_3(x,t,y,s) = y - x,$$

$$p_4(x,t,y,s) = \text{sgn}\,(s-t)$$

In order to obtain estimates for an operator $[p]$, it is necessary to have a more-or less closed form expression for Ω_i. This is the content of the next lemma.

Lemma 3.1.
(a) If $i = 1$, or if $i > 1$ and $\eta_j = (\psi_j, +)$ for $1 \le j \le i$, then

$$\Omega_i(x,y,u) = [sgn\,(y-x)]^i \int_{|x-y|}^\infty \int_{\xi_i}^\infty \cdots \int_{\xi_2}^\infty L_i(x, y, sgn\,(y-x)\xi_1, ..., sgn\,(y-x)\xi_i\,)\,\Omega_0(\xi_1, 0, u)d\xi$$

where $d\xi = d\xi_1 d\xi_2 \cdots d\xi_i$ and $L_i(x,y,\xi_1,...,\xi_i) = \prod_{j=1}^{i} \psi_j(x+\xi_j)$.

(b) Suppose that $i + 1 = j_0 > j_1 > j_2 > \cdots > j_k > j_{k+1} = 1$, where $j_1, ..., j_k$ are

precisely those integers $j \in [2, i]$ *for which* $\eta_j = (\psi_j, -)$. *Define*

$$\rho = \sum_{l=1}^{[\frac{k+1}{2}]} (j_{(2l-1)} - j_{2l}) ,$$

$$H_{r+1}(x, y, \xi_1, ..., \xi_i) = \prod_{j=j_{r+1}}^{j_r - 1} \psi_j(w + (-1)^r \xi_j), \ w = \begin{cases} y, & r \text{ odd} \\ x, & r \text{ even} \end{cases} \quad 0 \le r \le k;$$

$$L = \prod_{r=0}^{k} H_{r+1}.$$

Then

$$\Omega_i(x, y, u) = (-1)^\rho [sgn\,(y - x)]^i \int_{\tau_i}^{\infty} \cdots \int_{\tau_1}^{\infty} L(x, y, sgn\,(y - x)\xi_1, \cdots,$$

$$sgn\,(y - x)\xi_i)\Omega_0(y - sgn\,(y - x)\xi_{j_k}, x + sgn\,(y - x)\xi_1, u)d\xi,$$

where

(3.1)
$$\begin{cases} \tau_{j_0 - 1} = \tau_{j_1 - 1} = \tau_i = |y - x|; \\ \tau_{j_r - 1} = \xi_{j_{(r-1)}} \text{ provided that } k \ge 2 \text{ and } 2 \le r \le k; \\ \text{if } 0 \le r \le k \text{ and } j_{r+1} \le j_r - 2, \text{ then } \tau_j = \xi_{j+1} \text{ for } j_r - 2 \ge j \ge j_{r+1}. \end{cases}$$

The proof of Lemma 3.1 is a tedious but straightforward induction, so we omit it. We note that we could have lumped Lemma 3.1(a) and (b) together. However, Lemma 3.1 will be used in the proof of Lemma 3.2 which is somewhat technical. Therefore, to simplify matters for the reader we shall often first consider the simpler case, (a), of Lemma 3.1. Before stating Lemma 3.2 we need one last bit of notation. Let

(3.2) $$A(0) = 1; \ A(1) = a_1 + a_2; \ A(l) = a_1^{l-2}(a_1^2 + a_2^2) \text{ for } l \ge 2.$$

Then we have

Lemma 3.2. *There is a constant* $c_3 > 0$, *depending upon* n *and* α, *such that the following estimates hold:*
(a) If i, l *are nonnegative integers for which* $i + l$ *is odd , and if* F *is one of the representatives* $p_1^l, p_2^l, p_4 p_1^l, p_4 p_2^l$, *then for* $g \in C_0^\infty(R^2)$,

(3.3) $$\max\{||[F]g||_2, \ ||[F]^* g||_2\} \le \beta_1(c_3) A(l) \delta_i^{-1} ||g||_2 ;$$

moreover, if $g \in L^2(R^2) \cap L^\infty(R^2)$ *or* $g \equiv 1$, *we have*

(3.4) $$\max\{||[F]g||_{*,\alpha}, \ ||[F]^* g||_{*,\alpha}\} \le \beta_1(c_3) A(l) \delta_i^{-1} ||g||_\infty.$$

(b) If i, l are nonnegative integers, $l \geq 1$, and $i + l$ is odd, and if F is one of the representatives $p_3 p_1^{l-1}, p_3 p_2^{l-1}, p_3 p_4 p_1^{l-1}, p_3 p_4 p_2^{l-1}$, then for $g \in C_0^\infty(R^2)$,

$$\max\{\|[F]g\|_2, \|[F]^* g\|_2\} \leq \beta_1(c_3) A(l-1)\delta_i^{-1}\|g\|_2; \tag{3.5}$$

moreover, if $g \in L^2(R^2) \cap L^\infty(R^2)$ or $g \equiv 1$, we have

$$\max\{\|[F]g\|_{*,\alpha}, \|[F]^* g\|_{*,\alpha}\} \leq \beta_1(c_3) A(l-1)\delta_i^{-1}\|g\|_\infty. \tag{3.6}$$

Proof. To prove Lemma 3.2 we use yet again, the T1 Theorem. We prove (1.9) and (1.10) directly, while to prove (1.11) we use an induction argument. To begin, note that if $i = 1$, or if $i > 1$ and $\eta_j = (\psi_j, +)$ for $1 \leq j \leq i$, then Lemma 3.1(a) yields

(3.7)

$$|\Omega_i(x,y,u) - (-1)^i \Omega_i(y,x,u)| \leq c^{i+1} \sum_{j=1}^{i} (\hat{\delta}_{i,j})^{-1} M_j(x,y,u), \ (x,y,u) \in R^3,$$

where, for $1 \leq j \leq i$,

$$M_j(x,y,u)$$

$$= \int_{|x-y|}^\infty \int_{\xi_i}^\infty \cdots \int_{\xi_2}^\infty |\psi_j(x + \text{sgn}\,(y-x)\xi_j) - \psi_j(y - \text{sgn}\,(y-x)\xi_j)| \cdot \Omega_0(\xi_1, 0, u) d\xi. \tag{3.8}$$

Clearly, $M_j \equiv 0$ if $\psi_j \equiv 1$. Moreover, the constant c appearing on the right-hand side of (3.7) depends only upon the L^∞ norm of ϕ.

Now let $\bar{\sigma} = \bar{\sigma}(i,l) = (n - 1 + i + l)\alpha + 1 = \sigma(i,l) - 2\alpha$, where σ is as in section 2 and for fixed $x, y, \in R$, let χ_j be the characteristic function of the support of the function

$$G(\xi) = \psi_j(x + \text{sgn}\,(y-x)\xi) - \psi_j(y - \text{sgn}\,(y-x)\xi), \ \xi \in R.$$

Suppose first that $j = 1$. Then, letting $\lambda = -l$, $\nu = \sigma(0,l) - \bar{\sigma}(i,l) = (2-i)\alpha$, and using (3.8), (2.1a), (2.4), and the definition of Ω_0, we have

$$\int_R |u|^{-\bar{\sigma}} M_1(x,y,u) du$$

$$\leq c \int_{|x-y|}^\infty \int_{\xi_i}^\infty \cdots \int_{\xi_2}^\infty |G(\xi_1)| \int_R |\xi_1|^\lambda |u|^\nu V_{0,l}(\xi_1, u) du d\xi \tag{3.9a}$$

$$\leq c\Gamma(\bar{\sigma} - 1) \int_{|x-y|}^\infty \int_{\xi_i}^\infty \cdots \int_{\xi_2}^\infty |G(\xi_1)| \xi_1^{-l-i-1}\, d\xi.$$

On the other hand if $j \geq 2$, we let $\lambda = -1$,

$$\nu = \sigma(j-1,l) - \bar{\sigma}(i,l) = (l + n + j)\alpha - (n - 1 + i + l)\alpha = (j + 1 - i)\alpha$$

and use (2.1)(c) and (2.4) to obtain as in (3.9)(a)

(3.9b)

$$\int_{\mathbf{R}} |u|^{-\bar{\sigma}} M_j(x,y,u)du$$

$$\leq c\int_{|x-y|}^{\infty}\int_{\xi_i}^{\infty}\cdots\int_{\xi_{j+1}}^{\infty}|G(\xi_j)|\,(\int_{\mathbf{R}}|u|^{\nu}|\xi_j|^{\lambda}V_{j-1,1}(\xi_j,u)du)d\xi_j\cdots d\xi_i$$

$$\leq c\frac{\Gamma(\bar{\sigma}-1)\Gamma(i-j+l+2)}{\Gamma(i+l+1)}\int_{|x-y|}^{\infty}\cdots\int_{\xi_{j+1}}^{\infty}|G(\xi_j)|\xi_j^{-(l+i-j+2)}d\xi_j\cdots d\xi_i.$$

Comparing (3.9a) and (3.9b), we see that (3.9b) is valid for $j \geq 1$. We write

$$\int_{|x-y|}^{\infty}\cdots\int_{\xi_{j+1}}^{\infty}|G(\xi_j)|\xi_j^{-(l+i-j+2)}d\xi_j\cdots d\xi_i$$

(3.10)
$$= \int_{|x-y|}^{\zeta_j}\cdots\int_{\xi_{j+1}}^{\infty}\cdots + \int_{\zeta_j}^{\infty}\cdots\int_{\xi_{j+1}}^{\infty}\cdots$$

$$= J_1(x,y) + J_2(x,y),$$

where $\zeta_j = \max\{\mu_j, |x-y|\}$. To estimate J_1 we note that $J_1 = 0$ when $\mu_j \leq |x-y|$. Otherwise if $\xi_j \geq |x-y|$, then

(+)
$$|G(\xi_j)| \leq c\mu_j^{-2}(|x-y| + |\xi_j|) \leq c\,\mu_j^{-2}\,\xi_j\,.$$

Since $\xi_j \geq |x-y|$ in the integral defining J_1, it follows from (+) and integration that

$$|J_1| \leq c\mu_j^{-2}\,\frac{\Gamma(l+1)}{\Gamma(i-j+l+1)}\,|x-y|^{-l}(1\,+\,\log(\frac{\mu_j}{|x-y|}\,)),$$

where we really only need the logarithm term when $l = 0$. To estimate J_2 we note that the support of G has measure at most $c\mu_j$ and

(++)
$$||G||_{\infty} \leq c\mu_j^{-1}\,.$$

Using these estimates in the integral defining J_2 we deduce for $j+2 < i$,

$$|J_2| \leq c\mu_j^{-1}\int_{\zeta_j}^{\infty}\cdots\int_{\xi_{j+1}}^{\infty}\xi_j^{-(i-j+l+2)}\chi_j(\xi_j)d\xi_j\cdots d\xi_{i-1}d\xi_i$$

$$\leq c\int_{\zeta_j}^{\infty}\cdots\int_{\xi_{j+2}}^{\infty}\xi_{j+1}^{-(i-j+l+2)}d\xi_{j+1}\cdots d\xi_{i-1}d\xi_i$$

$$\leq c\frac{\Gamma(l+2)}{\Gamma(i-j+l+2)}\,\zeta_j^{-l-2}\,.$$

A similar estimate holds when $j+2 \geq i$. Using these estimates for J_1, J_2, and (3.9b), (3.10), we conclude

(3.11)
$$\int_{\mathbf{R}} |u|^{-\bar{\sigma}} M_j(x,y,u)du \leq \beta_1(c) \begin{cases} \mu_j^{-2}|x-y|^{-l}(1+\log(\frac{\mu_j}{|x-y|})), & |x-y| \leq \mu_j \\ |x-y|^{-l-2}, & |x-y| > \mu_j \end{cases}.$$

Next, suppose we are in the situation of Lemma 3.1(b). We once again obtain (3.7), but now

(3.12)
$$M_j(x,y,u) = \int_{\tau_i}^{\infty} \cdots \int_{\tau_1}^{\infty} |G(\xi_j)|\, \Omega_0(y - \text{sgn}\,(y-x)\xi_{j_k}, x + \text{sgn}\,(y-x)\xi_1, u)d\xi$$

for $1 \leq j \leq i$, where $\tau_1, ..., \tau_i$ are as in (3.1). As before, $M_j \equiv 0$ if $\psi_j \equiv 1$. Let r be the integer in $[1,k]$ such that $j_{r+1} \leq j \leq j_r - 1$. Note from Lemma 3.1(b) that $\tau_m \geq \tau_i = |x-y|$, for $1 \leq m \leq i$. If $\xi_j, \xi_{j_r} \geq \tau_i$, then clearly

$$\max\{\xi_j, \tau_i\} \leq \xi_j + \xi_{j_r} - \tau_i .$$

From these observations we find for ξ_j, ξ_{j_r} as above that

(+++)
$$|G(\xi_j)| \leq c\mu_j^{-2}(\xi_j + \xi_{j_r} - \tau_i),$$

We put

(3.13)
$$\int_{\mathbf{R}} |u|^{-\bar{\sigma}} M_j(x,y,u)du = J_3(x,y) + J_4(x,y)\,,$$

where

$$J_3(x,y) = \int_{\mathbf{R}} |u|^{-\bar{\sigma}} \int_{\tau_i}^{\zeta_j} \int_{\tau_{i-l}}^{\infty} \cdots \int_{\tau_1}^{\infty} |G(\xi_j)| \cdot \Omega_0(y - \text{sgn}(y-x)\xi_{j_k},\ x + \text{sgn}(y-x)\xi_1, u)d\xi du$$

$$J_4(x,y) = \int_{\mathbf{R}} |u|^{-\bar{\sigma}} \int_{\zeta_j}^{\infty} \int_{\tau_{i-1}}^{\infty} \cdots \int_{\tau_1}^{\infty} |G(\xi_j)| \cdot \Omega_0(y - \text{sgn}(y-x)\xi_{j_k},\ x + \text{sgn}(y-x)\xi_1, u)d\xi du.$$

Again $\zeta_j = \max\{|x-y|, \mu_j\}$, and $J_3 = 0$, unless $|x-y| < \mu_j$. In this case we see from (+++), as in (3.9b) and the estimate involving J_1, that

(3.14)
$$J_3(x,y) \leq c\mu_j^{-2} \int_{\tau_i}^{\zeta_j} \int_{\tau_{i-1}}^{\infty} \cdots \int_{\tau_1}^{\infty} (\xi_j + \xi_{j_r} - \tau_i) \int_{\mathbf{R}} (\xi_1 + \xi_{j_k} - \tau_i)^{\lambda} |u|^{\nu} \cdot V_{0,l}(\xi_1 + \xi_{j_k} - \tau_i, u)du d\xi$$

$$\leq c\mu_j^{-2}\Gamma(\bar{\sigma}-1) \int_{\tau_i}^{\zeta_j} \int_{\tau_{i-1}}^{\infty} \cdots \int_{\tau_1}^{\infty} (\xi_j + \xi_{j_r} - \tau_i)(\xi_1 + \xi_{j_k} - \tau_i)^{-l-i-1} d\xi\,.$$

Here we have also used (2.1)(a) and (2.4) with $\lambda = -l$, $\nu = (2-i)\alpha$. Subsequent integration of (3.14) yields

$$J_3(x,y) \le c\mu_j^{-2} \frac{\Gamma(\bar\sigma - 1)\Gamma(i-j+l+2)}{\Gamma(i+l+1)} \int_{\tau_i}^{\zeta_j} \cdots \int_{\tau_j}^{\infty}$$

(3.15)
$$(\xi_j + \xi_{j_r} - \tau_i)^{-(l+i-j+1)} d\xi_j \cdots d\xi_i$$

$$\le \beta_1(c)\mu_j^{-2}|x-y|^{-l}\left(1 + \log(\tfrac{\mu_j}{|x-y|})\right).$$

To estimate J_4, we again note that the support of G has measure at most $c\mu_j$. Using this fact and (++) we obtain as in (3.14), (3.15), and and the estimate of J_2 that

(3.16)
$$J_4(x,y)$$

$$\le c\mu_j^{-1} \frac{\Gamma(\bar\sigma - 1)\Gamma(i-j+l+2)}{\Gamma(i+l+1)} \int_{\zeta_j}^{\infty}\int_{\tau_{i-1}}^{\infty} \cdots \int_{\tau_j}^{\infty} (\xi_j + \xi_{j_r} - \tau_i)^{-(l+i-j+2)}$$

$$\cdot \chi_j(\xi_j) d\xi_j \cdots d\xi_{i-1} d\xi_i$$

$$\le c\, \frac{\Gamma(\bar\sigma - 1)\Gamma(i-j+l+2)}{\Gamma(i+l+1)} \int_{\zeta_j}^{\infty}\int_{\tau_{i-1}}^{\infty} \cdots \int_{\tau_{j+1}}^{\infty} (\tau_j + \xi_{j_r} - \tau_i)^{-(l+i-j+2)}$$

$$d\xi_{j+1} \cdots d\xi_{i-1} d\xi_i$$

$$\le \beta_1(c) \int_{\zeta_j}^{\infty} \xi_i^{-l-3} d\xi_i \le \beta_1(c)\, \zeta_j^{-l-2}.$$

From (3.16), (3.15), and (3.13), we see that (3.11) also holds when Ω_i is as in Lemma 3.1 (b) .

To continue the proof of Lemma 3.2 note that, by the recursion relation defining Ω_i, we have

(3.17) $$|\Omega_i(x,y,u)| \le c^{i+1}\delta_i^{-1}|u|^{\sigma}|x-y|^{-l} V_{i,l}(x-y,u), \ i \in \{0\} \cup N.$$

Suppose, until further notice, that either (i) $l \ge 1$ and we are in the situation of either Lemma 3.2(a) or (b); or (ii) $l = 0$ and we are in the situation of Lemma 3.2 (a) . Let us write

$$\Psi_1(x,t,y,s) = |s-t|^{-\bar\sigma} F(x,t,y,s)\Omega_i(x,y,s-t)$$

so that

(3.18) $$[F]g(x,t) = \iint_{R^2} \Psi_1(x,t,y,s) g(y,s) dy ds, \ (x,t) \in R^2, \ g \in C_0^\infty(R^2).$$

We resume the notation used in the proof of Proposition 2.5, to wit: let $d > 0$, $\lambda = d/2$, $K = 2\cdot 6^{1/\alpha^2}$, $\tilde Q = Q_{Kd}(x_0,t_0)$; let $\theta, E_j, \mathcal{H}_j$ $(j = 1,2,3)$ be as defined therein. As in the discussion preceding (2.34), let $P^\alpha = (K^\alpha+1)2^{-\alpha}$, $Q^\sharp =$

$Q_{2Pd}(x_0,t_0)$. If $|x_0-y_0|\le (K/2)^\alpha d^\alpha$ and $|t_0-s_0|<Kd/2$, then the support of $\theta_\lambda^{(y_0,s_0)}$ is in $Q^\sharp$, and we have in view of (3.18) ,

$$\langle (D\theta)_\lambda^{(x_0,t_0)}, [F](\theta_\lambda^{(y_0,t_0)})\rangle = (J_5+J_6)(x_0,t_0,y_0,s_0), \tag{3.19}$$

where

$$J_5(x_0,t_0,y_0,s_0) = \int_{R^2}\int_{R^2} \Psi_1(x,t,y,s)H(x,t,y,s)dydsdxdt, \tag{3.20a}$$

(3.20b)

$$H(x,t,y,s)=\frac{1}{2}[(D\theta)_\lambda^{(x_0,t_0)}(x,t)\theta_\lambda^{(y_0,t_0)}(y,s)-(D\theta)_\lambda^{(x_0,t_0)}(y,t)\theta_\lambda^{(y_0,t_0)}(x,s)],$$

(3.20c)

$$J_6(x_0,t_0,y_0,s_0)=\int_{R^2}\int_{R^2}\tfrac{1}{2}[\Psi_1(x,t,y,s)+\Psi_1(y,t,x,s)](D\theta)_\lambda^{(x_0,t_0)}(y,t) \cdot \theta_\lambda^{(y_0,s_0)}(x,s)dydsdxdt.$$

The support of H is contained in $Q^\sharp\times Q^\sharp$, and we have

$$|H(x,t,y,s)|\le c|x-y|\lambda^{-2-3\alpha}, \tag{3.21}$$

so that, by (3.17), (3.20)(a), (3.21), together with (0.1), (0.4), (3.2), and (2.18a),

(3.22)

$$\begin{aligned}|J_5(\ x_0,t_0,y_0,s_0)| &\le c\lambda^{-1-2\alpha}\sup_{(x,t)\in Q^\sharp}\iint_{Q^\sharp}|\Psi_1(x,t,y,s)|\,|x-y|dyds\\
&\le c^{i+1}\bar A\delta_i^{-1}\lambda^{-1-2\alpha}\sup_{(x,t)\in Q^\sharp}\iint_{Q^\sharp}|s-t|^{2\alpha}|x-y|V_{i,l}(x-y,s-t)dyds\\
&\le c^{i+1}\bar A\delta_i^{-1}\lambda^{2\alpha}\iint_{\substack{|z|\le 4P^\alpha\\ |u|\le 4P}}|u|^{2\alpha}|z|V_{i,l}(\lambda^\alpha z,\lambda u)dzdu\\
&\le \bar A\delta_i^{-1}\lambda^{-1-\alpha}\beta_1(c)\ ,\end{aligned}$$

where the last inequality follows from (2.3) with $\lambda=1$ and $\nu=2\alpha$. Also, $\bar A=A(l)$ in the situation of Lemma 3.2 (a) and $\bar A=A(l-1)$ in the situation of Lemma 3.2 (b) . As in section 2 let $\mathcal{H}_0$ be the characteristic function of $Q_{2K}(0,0)$; then (3.22) implies

$$|J_5(x_0,t_0,y_0,s_0)|\le \bar A\delta_i^{-1}\beta_1(c)(\mathcal{H}_0)_\lambda(x_0-y_0,s_0-t_0). \tag{3.23}$$

To estimate J_6, recall that $i+l$ is odd; if $i=0$, we easily see that $J_6\equiv 0$. Otherwise, note that if F is one of the representatives in Lemma 3.2 (a) or (b), then

$$F(x,t,y,s) = (-1)^l F(y,t,x,s) = -(-1)^i F(y,t,x,s),$$

whereupon (3.20c), (0.1), (3.7), and (3.11) yield

(3.24)

$$\lambda^{1+\alpha}|J_6(x_0,t_0,y_0,s_0)| \le c \sup_{(x,t)\in Q^\sharp} \iint\limits_{Q^\sharp} |\Psi_1(x,t,y,s)+\Psi_1(y,s,x,t)|dyds$$

$$\le c \sup_{(x,t)\in Q^\sharp} \iint\limits_{Q^\sharp} |s-t|^{-\bar{\sigma}}|F(x,t,y,s)|$$

$$\cdot |\Omega_i(x,y,s-t)+(-1)^i\Omega(y,x,s-t)|dyds$$

$$\le c^{i+1}\bar{A} \sup_{(x,t)\in Q^\sharp} \sum_{j=1}^{i} (\hat{\delta}_{i,j})^{-1} \int_{\{|x-y|\le 4P^\alpha d^\alpha\}} |x-y|^l \int_{\boldsymbol{R}} |u|^{-\bar{\sigma}} M_j(x,y,u)dudy$$

$$\le \bar{A}\beta_1(c)\sum_{j=1}^{i} \{\mu_j^{-1}\delta_i^{-1} \int_0^{\mu_j} (1+\log(\tfrac{\mu_j}{z}))\, dz + (\hat{\delta}_{i,j})^{-1} \int_{\mu_j}^{\infty} z^{-2}\}dz$$

$$\le \bar{A}\beta_1(c)\delta_i^{-1}$$

so that

$$|J_6(x_0,t_0,y_0,s_0)| \le \bar{A}\delta_i^{-1}\beta_1(c)(\mathcal{H}_0)_\lambda(x_0-y_o,s_0-t_0). \tag{3.25}$$

Combining (3.19), (3.23), and (3.25) yields

$$\begin{aligned} &|\langle (D\theta)_\lambda^{(x_0,t_0)},\ [F](\theta_\lambda^{(y_0,t_0)})\rangle| \\ &\le \bar{A}\delta_i^{-1}\beta_1(c)(\mathcal{H}_0)_\lambda(x_0-y_0,s_0-t_0), \end{aligned} \tag{3.26}$$

when $|x_0-y_0| < (K/2)^\alpha d^\alpha$, and $|t_0-s_0| < Kd/2$. If x_0, y_0, s_0, t_0 are not as in (3.26), then, for some $j \in \{1,2,3\}$ we have

$$\operatorname{supp}(\theta_\lambda^{(y_0,t_0)}) \subset E_j \tag{3.27}$$

(compare (2.35) and following). If (3.27) holds for $j = 1$ or 2, then

(3.28)

$$|\langle (D\theta)_\lambda^{(x_0,t_0)},[F](\theta_\lambda^{(y_0,t_0)})\rangle|$$

$$\le c^{i+1}\bar{A}\delta_i^{-1}\lambda^{-1-\alpha} \sup_{(x,t)\in Q} \iint\limits_{E_j\cap Q_d(y_0,t_0)} |s-t|^{2\alpha}V_{i,l}(x-y,s-t)dyds,$$

by (3.17) and (3.18). Exactly as in (2.37)-(2.38), we obtain

$$|\langle (D\theta)_\lambda^{(x_0,t_0)},[F](\theta_\lambda^{(y_0,t_0)})\rangle| \le c^{i+1}\bar{A}\delta_i^{-1}(\mathcal{H}_j)_\lambda(x_0-y_0,s_0-t_0) \tag{3.29}$$

where $\mathcal{H}_1$ and $\mathcal{H}_2$ are the same as in the earlier estimates. Finally, as in (2.39)-(2.40) we see that, if (3.27) holds for $j = 3$, then

(3.30)
$$|\langle (D\theta)_\lambda^{(x_0,t_0)}, [F](\theta_\lambda^{(y_0,s_0)})\rangle| \leq c\lambda^{-1-\alpha} \iint_{E_3 \cap Q_d(y_0,t_0)} |\Psi_1(x,t,y,s) - \Psi_1(x_0,t_0,y_0,s_0)| dyds\,.$$

To estimate the integrand in (3.30) we note that

$$|\frac{\partial}{\partial u}\Omega_1(x,y,u)| \leq c\mu_1^{-1} \int_{|x-y|}^{\infty} \xi^{n-2+1/\alpha}|u|^{-2} \exp\left[-|\xi|^{\frac{1}{\alpha}}\,|u|^{-1}\right] du\,.$$

Also if $j > 1$, then

$$|\frac{\partial}{\partial u}\Omega_j(x,y,u)| \leq c\mu_j^{-1} \int_y^{\infty} |\frac{\partial}{\partial u}\Omega_{j-1}(x,\xi,u)|\, d\xi\,,$$

when $\eta_j = (\psi_j, +)$. If $\eta_j = (\psi_j, -)$, interchange x and ξ in the arguments of Ω_{j-1}. Using these notes, induction, and (3.17), we deduce for $i \geq 1$

(3.31a)
$$\left|\frac{\partial}{\partial u}(|u|^{-\bar{\sigma}}\Omega_i(x,y,u))\right| \leq c^{i+1}\delta_i^{-1}\{\bar{\sigma}|u|^{2\alpha-1}|x-y|^{-l}\, V_{i,l}(x-y,u) + |u|^{-\bar{\sigma}-2} \int_{|x-y|}^{\infty} \int_{\xi_i}^{\infty} \cdots \int_{\xi_2}^{\infty} |\xi_1|^{1/\alpha}\Omega_0(\xi_1,0,u)d\xi\}.$$

In addition, using the recursion relation defining Ω_i as in (3.31a), we obtain

(3.32a)
$$\max\left\{\left|\frac{\partial}{\partial x}\,\Omega_i(x,y,u)\right|, \left|\frac{\partial}{\partial y}\,\Omega_i(x,y,u)\right|\right\} \leq c^{i+1}\delta_i^{-1}|x-y|^{-l}|u|^{\sigma(i,l)-\alpha}\, V_{(i-1),l}(x-y,u).$$

When $i = 0$,

(3.31b)
$$\left|\frac{\partial}{\partial u}\,(|u|^{-\bar{\sigma}}\Omega_i(x,y,u))\right| \leq c\bar{\sigma}|u|^{2\alpha-1}|x-y|^{-l}(1+|x-y|^{1/\alpha}|u|^{-1})V_{0,l}(x-y,u)$$

and

(3.32b)
$$\max\left\{\left|\frac{\partial}{\partial x}\,\Omega_0(x,y,u)\right|, \left|\frac{\partial}{\partial y}\,\Omega_0(x,y,u)\right|\right\}$$
$$\leq c|x-y|^{n-2}\{|x-y|^{-1}(n-2+\alpha^{-1}|x-y|^{1/\alpha}|u|^{-1})+$$
$$\epsilon^{-1}|\gamma'(|x-y|\epsilon^{-1})| + N^{-1}|\gamma'(|x-y|N^{-1})|\}\exp\left(-|x-y|^{1/\alpha}|u|^{-1}\right).$$

By analogy to the derivation of (2.41), the estimates (3.31)-(3.32) yield

(3.33) $|\Psi_1(x,t,y,s) - \Psi_1(x_0,t_0,y_0,s_0)| \le \bar{A}\delta_i^{-1}c^{i+l+1}(\mathcal{H}_3)_\lambda(x_0 - y_0, s_0 - t_0)$

for $(x,t) \in Q$ and $(y,s) \in E_3 \cap Q_d(y_0,t_0)$; here $\mathcal{H}_3$ is defined by (2.19) with L, M, N as in (2.41), and we use the fact that sgn $(s-t) =$ sgn $(s_0 - t_0)$. Combining (3.30) and (3.33) yields

$$(3.34)\qquad |\langle (D\theta)_\lambda^{(x_0,t_0)}, [F](\theta_\lambda^{(y_0,s_0)})\rangle| \le \bar{A}\delta_i^{-1}c^{i+l+1}(\mathcal{H}_3)_\lambda(x_0 - y_0, s_0 - t_0).$$

Now let c be the largest constant appearing in (3.26), (3.29), and (3.34). Set

$$\mathcal{H} = \bar{A}\delta_i^{-1}[\beta_1(c)\mathcal{H}_0 + c^{i+l+1}(\mathcal{H}_1 + \mathcal{H}_2 + \mathcal{H}_3)]$$

and

$$a_3 = \iint\limits_{R^2} (|y| + |s|^\alpha + 1)^{1/2}\mathcal{H}(y,s)dyds.$$

Then (2.20) yields $a_3 \le \beta_1(c)\bar{A}\delta_i^{-1}$. Taking $h = a_3^{-1}\mathcal{H}$, we obtain (1.7)-(1.9) of the T1 Theorem with $[F]$ in place of T, $\delta = 1/2$, and a_3 as above. Applying a similar argument to $[F]^*$ yields (1.10) of the T1 Theorem with $[F]$ in place of T and $a_4 = a_3$.

It remains to establish (1.11) of the T1 Theorem for $T = [F]$. In fact we prove that

$$(3.35)\qquad \max\{||[F]1||_{*,\alpha},\ ||[F]^*1||_{*,\alpha}\} \le c_3^{-\frac{1}{2}}\beta_1(c_3)\bar{A}\delta_i^{-1},$$

for c_3 sufficiently large. We note that $\Psi_1(x,t,\cdot,\cdot)$ and $\Psi_1(\cdot,\cdot,x,t)$ are absolutely integrable, as follows easily from the definition of γ, Ω_0. Thus $[F]1$ and $[F]^*1$ are well-defined.

To begin, suppose that either (+) $l = 0$ and we are in the situation of Lemma 3.2(a) or (++) $l = 1$ and we are in the situation of Lemma 3.2(b). Write $\tilde{Q} = I_1 \times I_2$, as in section 2, so that

$$\begin{aligned}[F]1(x,t) &\int_{\boldsymbol{R}^2} \Psi_1(x,t,y,s)dyds \\ (3.36\text{a})\qquad &= \int_{\boldsymbol{R}}\int_x^\infty \{\Psi_1(x,t,y,s) + \Psi_1(x,t,2x-y,s)\}dyds \\ &= B_1(x,t) + B_2(x,t)\end{aligned}$$

where

$$(3.36\text{b})\quad B_1(x,t) = \int_{\boldsymbol{R}}\int_x^\infty \{\Psi_1(x,t,y,s) + \Psi_1(x,t,2x-y,s)\}\chi_{I_1}(y)dyds$$

$$(3.36\text{c})\quad B_2(x,t) = \int_{\boldsymbol{R}}\int_x^\infty \{\Psi_1(x,t,y,s) + \Psi_1(x,t,2x-y,s)\}\chi_{\boldsymbol{R}\setminus I_1}(y)dyds$$

for $(x,t) \in Q$. The strategy is to repeat the arguments following (3.7), with $\Omega_i(x, 2x-y, u)$ in place of $\Omega_i(y,x,u)$. By analogy to (3.7), we obtain, for $i \geq 1$,

$$(3.37) \quad |\Omega_i(x,y,u) - (-1)^i \Omega_i(x, 2x-y, u)| \leq c^{i+1} \sum_{j=1}^{i} (\hat{\delta}_{i,j})^{-1} \hat{M}_j(x,y,u),$$

where

$$(3.38a) \quad \begin{aligned} \hat{M}_j(x,y,u) = \int_{|x-y|}^{\infty} \int_{\xi_i}^{\infty} \cdots \int_{\xi_2}^{\infty} &\{|\psi_j(x \pm \xi_j) - \psi_j(x \mp \xi_j)| \\ &+ |\psi_j(y \mp \xi_j) - \psi_j(2x-y \pm \xi_j)|\} \Omega_0(\xi_1, 0, u) d\xi \end{aligned}$$

when $i = 1$, or $i \geq 1$ and $\eta_j = (\psi_j, +)$ for $1 \leq j \leq i$; while

$$(3.38b) \quad \begin{aligned} \hat{M}_j(x,y,u) = \int_{|x-y|}^{\infty} \int_{\tau_{i-1}}^{\infty} \cdots \int_{\tau_1}^{\infty} &\{|\psi_j(x \pm \xi_j) - \psi_j(x \mp \xi_j)| \\ &+ |\psi_j(y \mp \xi_j) - \psi_j(2x-y \pm \xi_j)|\} \Omega_0(y \mp \xi_{j_k}, x \pm \xi_1, u) d\xi \end{aligned}$$

otherwise . It is easily seen that (3.11) holds with $\hat{M}_j$ in place of M_j. As in the estimate (3.25), we deduce that

$$(3.39) \quad |B_1(x,t)| \leq \bar{A} \delta_i^{-1} \beta_1(c), \ (x,t) \in Q.$$

Arguing as in (3.28), (3.29), and (3.34), we obtain

$$(3.40) \quad |B_2(x,t) - B_2(z,u)| \leq \bar{A} \delta_i^{-1} \beta_1(c), \ (x,t), (z,u) \in Q.$$

Combining (3.36a), (3.39)-(3.40), it follows as in the derivation of (2.33) that the estimate on $[F]1$ in (3.35) is valid for c_3 sufficiently large when $i \geq 1$ and either $l = 0$ (Lemma 3.2(a)) or $l = 1$ (Lemma 3.2(b)). If $i = 0$, $l = 1$ and we are in the situation of Lemma 3.2(b), then $[F]1 \equiv 0$ and hence trivially satisfies the desired estimates. Analogous considerations for $[F]^*$ yield (3.35) in these specific cases. Moreover, it is easily verified that in these cases, $[F]$ maps $C_0^\infty(R^2)$ boundedly into $L^2(R^2)$. Since we have already shown that (1.9)-(1.10) hold for $[F]$, we conclude from the T1 Theorem and the argument of Peetre, as in the conclusion to the proof of Proposition 2.5, that (3.3)-(3.6) are true when either $l = 0$ and $[F]$ is as in Lemma 3.2(a) , or $l = 1$ and $[F]$ is as in Lemma 3.2(b).

To establish (3.35) in full generality, we proceed by induction from the basic cases considered above. The induction depends on the introduction of yet another operator, this one on $L^2(R)$. Suppose that G is one of the following representatives: p_2^l, $p_4 p_2^l$, $p_3 p_2^{l-1}$, $p_3 p_4 p_2^{l-1}$. For fixed $t \in R$ and i, l, ϵ, N, as in lemma 3.2, define

$$(3.41) \quad \begin{aligned} \{G\}_t k(x) &= \{G\} k(x) \\ &= \iint_{R^2} |s-t|^{-\bar{\sigma}} G(x,t,y,s) \Omega_i(x,y,s-t) k(y) dy ds \end{aligned}$$

This definition makes sense because G is independent of s. Slight modifications of our previous arguments yield the estimate

$$(3.42)\qquad |\langle(\phi')_\lambda^{x_0}, \{G\}_t(\phi_\lambda^{y_0})\rangle| \le \bar{A}\delta_i^{-1}\beta_1(c)h_\lambda(x_0 - y_0)$$

where the inner product is the ordinary one on $L^2(R), \phi$ is as previously defined and ϕ' is its derivative, h is a function on R satisfying

$$\int_R h(z)(1+|z|^{1/2})dz = 1,$$

$h_\lambda = \lambda^{-\alpha}h(\cdot\lambda^{-\alpha})$, and $\phi_\lambda^{(y_0)}(x) = \phi_\lambda(y_0 - x)$. The estimate (3.42) is a one dimensional version of (1.9), and can easily be seen to hold with the adjoint $\{G\}_t^*$ in place of $\{G\}_t$. The constant c in each case does not depend upon t.

In particular when $G = 1, p_3, p_4$, or p_3p_4, then from (3.36), (3.39), and (3.40) we see that G_t1 and G_t^*1 are in $BMO(R)$ for each fixed t with

$$(3.43)\qquad \max\{||\{G\}_t1||_*,\ ||\{G\}_t^*1||_*\} \le \bar{A}\delta_i^{-1}\beta_1(c).$$

(Here , $||.||_*$ denotes the ordinary BMO-norm on R to which 'nonisotropic' BMO-norms are equivalent.) As G_t maps $C_0^\infty(R)$ into $L^2(R)$, the one-dimensional T1 Theorem together with (3.42)-(3.43) imply that $\{G\}_t, \{G\}_t^*$ extend (for these specific choices of G) to bounded operators on $L^2(R)$ satisfying

$$(3.44)\qquad \max\{||\{G\}_tk||_2,\ ||\{G\}_t^*k||_2\} \le \bar{A}\delta_i^{-1}\beta_1(c^+)||k||_2,\ k \in L^2(R);$$

$$(3.45)\qquad \max\{||\{G\}_tk||_*,\ ||\{G\}_t^*||_*\} \le \bar{A}\delta_i^{-1}\beta_1(c^+)||k||_\infty,$$

when $k \in L^2(R) \cap L^\infty(R)$ or $k \equiv 1$, provided c^+ is large enough.

Now the induction begins. Suppose (3.35) and the estimates (3.44)-(3.45) (for $G = p_2^l, p_4p_2^l, p_3p_2^{l-1}, p_3p_4p_2^{l-1}$) hold for pairs (i,l) of nonnegative integers with $l \le m$ and $i+l$ odd. Then since we have already proved (1.9)-(1.10), it follows from the $T1$ Theorem that Lemma 3.2 is true for pairs (i,l) of nonegative integers with $l \le m$ and $i+l$ odd provided c_3 is large enough. Let $q'(y,s) = \frac{\partial q}{\partial y}(y,s)$, $\dot{q}(y,t) = \frac{\partial q}{\partial y}(y,t)$ for $y, s \in R$. Define two sequences $\bar{\eta}$ and $\tilde{\eta}$ based on η by setting

$$(3.46)\qquad \bar{\eta}_j = \begin{cases} \eta_j,\ j \ne i+1 \\ (1,+),\ j = i+1 \end{cases},\ \tilde{\eta}_j = \begin{cases} \eta_j,\ j \ne i+1 \\ (1,-),\ j = i+1 \end{cases}.$$

To avoid confusion, we indicate the dependence of operators of form $[F]$ on i and η, and of operators of form $\{G\}$ on i, η, t. Integration by parts yields the following identities for $l = m+1$ at (x,t).

$$
\begin{aligned}
&[p_1^{m+1}]_{i,\eta}1 = (m+1)[p_1^m]_{i+1,\bar\eta}(q') \\
&[p_4p_1^{m+1}]_{i,\eta}1 = (m+1)[p_4p_1^m]_{i+1,\bar\eta}(q') \\
&[p_2^{m+1}]^*_{i,\eta}1 = (m+1)(-1)^{m+1}[p_1^m]_{i+1,\bar\eta}(q') \\
&[p_4p_2^{m+1}]^*_{i,\eta}1 = (m+1)(-1)^{m+1}[p_1^m]_{i+1,\bar\eta}(q') \\
&[p_3p_1^m]_{i,\eta}1 = [p_1^m]_{i+1,\bar\eta}1 + m[p_3p_1^{m-1}]_{i+1,\bar\eta}(q') \\
&[p_3p_4p_1^m]_{i,\eta}1 = [p_4p_1^m]_{i+1,\bar\eta}1 + m[p_3p_4p_1^{m-1}]_{i+1,\bar\eta}(q') \\
&[p_3p_2^m]^*_{i,\eta}1 = (-1)^{m+1}[p_1^m]_{i+1,\bar\eta}1 + (-1)^{m+1}m[p_3p_1^{m-1}]_{i+1,\bar\eta}(q') \\
&[p_4p_3p_2^m]^*_{i,\eta}1 = (-1)^{m+1}[p_4p_1^m]_{i+1,\bar\eta}1 + m(-1)^m[p_4p_3p_1^{m-1}]_{i+1,\bar\eta}(q').
\end{aligned}
\tag{3.47}
$$

As mentioned above, the induction hypothesis implies that Lemma 3.2 holds for each of the operators on the right-hand side of the equalities in (3.47). Using (3.4) and (3.6) yields

$$\|T1\|_{*,\alpha} \le (1+m)\bar A(m)\beta_1(i+1,m,c_3)\delta_i^{-1} \tag{3.48}$$

whenever T is one of the operators on the left-hand side of (3.47). Integration by parts also yields, for $x,t \in R$, $l = m+1$:

(3.49)

$$
\begin{aligned}
&\{p_2^{m+1}\}_{i,\eta,t}1(x) = (m+1)\{p_2^m\}_{i,1,\bar\eta,t}(\dot q(\cdot,t))(x) \\
&\{p_4p_2^{m+1}\}_{i,\eta,t}1(x) = (m+1)\{p_4p_2^m\}_{i+1,\bar\eta,t}(\dot q(\cdot,t))(x) \\
&[p_1^{m+1}]^*_{i,\eta}1(x,t) = (m+1)(-1)^{m+1}\{p_2^m\}_{i+1,\bar\eta,t}(\dot q(\cdot,t))(x) \\
&[p_4p_1^{m+1}]^*_{i,\eta}1(x,t) = (m+1)(-1)^m\{p_2^m\}_{i+1,\bar\eta,t}(\dot q(\cdot,t))(x) \\
&\{p_3p_2^m\}_{i,\eta,t}1(x) = \{p_2^m\}_{i+1,\bar\eta,t}1(x) + m\{p_3p_2^{m-1}\}_{i+1,\bar\eta,t}(\dot q(\cdot,t))(x) \\
&\{p_3p_4p_2^m\}_{i,\eta,t}1(x) = \{p_4p_2^m\}_{i+1,\bar\eta,t}1(x) + m\{p_3p_4p_2^{m-1}\}_{i+1,\bar\eta,t}(\dot q(\cdot,t))(x) \\
&[p_3p_1^m]^*_{i,\eta}1(x,t) \\
&\qquad = (-1)^{m+1}\{p_2^m\}_{i+1,\bar\eta,t}1(x) + m(-1)^{m+1}\{p_3p_1^{m-1}\}_{i+1,\bar\eta,t}(\dot q(\cdot,t))(x) \\
&[p_4p_3p_1^m]^*_{i,\eta}1(x,t) \\
&\qquad = (-1)^m\{p_4p_2^m\}_{i+1,\bar\eta,t}1(x) + m(-1)^m\{p_4p_3p_2^{m-1}\}_{i+1,\bar\eta,t}(\dot q(\cdot,t))(x).
\end{aligned}
$$

Again the induction hypothesis applies to each operator on the right-hand side of (3.49); using (3.45) yields

$$\|\{G\}_{i,\eta,t}1\|_* \le 3(1+m)\bar A(m+1)\beta_1(i+1,m,c^+)\delta_i^{-1}\,, \tag{3.50}$$

for each $t \in R$, where $G \in \{p_2^{m+1}, p_4p_2^{m+1}, p_3p_2^m, p_3p_4p_2^m\}$. We also obtain (3.50) with $\{G\}^*$ in place of $\{G\}$, for the same choices of G, once we observe that

$$\begin{aligned}
&\{p_2^{m+1}\}^*_{i,\eta,t}1(x) = [p_1^{m+1}]^*_{i,\eta}1(x,t)\\
&\{p_4p_2^{m+1}\}^*_{i,\eta,t}1(x) = [p_4p_1^{m+1}]^*_{i,\eta}1(x,t)\\
&\{p_3p_2^{m}\}^*_{i,\eta,t}1(x) = [p_3p_1^{m}]^*_{i,\eta}1(x,t)\\
&\{p_3p_4p_2^{m}\}^*_{i,\eta,t}1(x) = [p_4p_3p_1^{m}]^*_{i,\eta}1(x,t).
\end{aligned} \tag{3.51}$$

Now from the definition of β_1 we observe that

$$mc^+\beta_1(i+1,m,c^+) \leq \beta_1(i,m+1,c^+)\,. \tag{3.52}$$

Using this observation, (3.49), (3.42), (3.50), (3.51),the one-dimensional T1 Theorem, and the Peetre-type argument, we find for c^+ large enough that (3.44)-(3.45) hold for the specified G's, for the pair $(i,m+1)$ of nonnegative integers.

Using the identities (3.51), the estimate (3.48), and (3.44) for the pair $(i,m+1)$, we claim that

$$\max\{||\mathrm{T}1||_{*,\alpha},\ ||\mathrm{T}^*1||_{*,\alpha}\} \leq c'm\bar{A}(m+1)\beta_1(i,m+1,c^+)\delta_i^{-1} \tag{3.53}$$

when $T=[F]_{i,\eta}$ for some $F\in\{p_1^{m+1},p_4p_1^{m+1},p_3p_1^m,p_4p_3p_1^m\}$. In (3.53), c' denotes a constant. Indeed, from (3.44), Peetre's argument, and earlier kernel estimates, we obtain, for $I=\{x\in R:\ |x-x_0|<(d/2)^\alpha\}$ and $\tilde{Q}=I_1\times I_2$ as before, that

$$2^{\alpha-1}d^{-\alpha}\int_I |\mathrm{T}1(x,t)-T\chi_{(\boldsymbol{R}\setminus I_1)}(x_0,t)|dx \leq c'\bar{A}(m+1)\beta_1(i,m+1,c^+)\delta_i^{-1}$$

By analogy to (3.33), we obtain

$$|T\chi_{\boldsymbol{R}\setminus I_1}(x_0,t)-T\chi_{\boldsymbol{R}\setminus I_1}(x_0,t_0)| \leq c'\bar{A}(m+1)\beta_1(i,m+1,c^+)\delta_i^{-1}$$

for $|t-t_0|<d/2$, whence

$$2^{\alpha-1}d^{-(1+\alpha)}\iint_Q |\mathrm{T}1(x,t)-T\chi_{\boldsymbol{R}\setminus I_1}(x_0,t_0)|dxdt \leq c'\bar{A}(m+1)\beta_1(i,m+1,c^+)\delta_i^{-1}.$$

As the same estimates hold for T^*, we obtain (3.53). From (3.48), observation (3.52) with c^+ replaced by c_3 , and (3.53) we see for c_3 large enough that (3.35) holds for $(i,m+1)$, and hence in full generality by induction. Lemma 3.2 now follows from the T1 Theorem and the usual Peetre-type argument. □

4. More Singular Integrals

We shall need a slightly more general lemma than Lemma 3.2. For this purpose we keep the notation in sections 2-3 and suppose r,k,l, are nonnegative integers with $k+r=l$. Recall that n,ϵ,N,η, are fixed and $l+i$ is odd. We prove

Lemma 4.1. *Replace c_3 in Lemma 3.2 by c_4. If c_4 is large enough and $k+r=l$, then (3.3), (3.4), are valid with p_1^l,p_2^l, replaced in (a) by $p_1^kp_2^r$, while if $k+r=l-1$, then (3.5), (3.6), remain true with p_1^{l-1},p_2^{l-1}, replaced in (b) by $p_1^kp_2^r$.*

Proof. We prove Lemma 4.1 only for $[p_1^k p_2^r]$ which is in the case $k + r = l$. The other proofs are similar. From Lemma 3.2 we see that Lemma 4.1 is valid when $l = 1$. Hence assume $l \geq 2$. We note that

$$p_1(x,t,y,s) - p_2(x,t,y,s) = q(y,s) - q(y,t) + q(x,t) - q(x,s).$$

From this note, (3.17), and Proposition (2.5) with l replaced by $l-2$, we deduce that (3.3)-(3.4) of Lemma 3.2 are valid whenever, $F = (p_1 - p_2)^2(p_1^{k_1}p_2^{r_1})$, and $k_1 + r_1 = l - 2$. Next we observe that the above representatives and p_1^l, p_2^l, can be written in matrix form as, MX, where

$$M = \begin{pmatrix} 1 & 0 & \cdot & \cdot & \cdot & \cdot & \cdot \\ 1 & -2 & 1 & 0 & \cdot & \cdot & \cdot \\ 0 & 1 & -2 & 1 & 0 & \cdot & \cdot \\ \cdot & \cdot & \cdot & \cdot & \cdot & \cdot & \cdot \\ 0 & \cdot & \cdot & \cdot & \cdot & \cdot & 1 \end{pmatrix}$$

$$X = \begin{pmatrix} p_1^l \\ p_1^{l-1}p_2 \\ \vdots \\ p_2^l \end{pmatrix}$$

From this observation, our deduction, and Lemma 3.2 we see that in order to prove Lemma 4.1 for $[p_1^k p_2^r]$, it suffices to estimate the coefficients of M^{-1}. Indeed this matrix problem is equivalent to showing $p_1^k p_2^r$, $k+r = l$, can be expressed as a linear combination of the above representatives, with estimates on the coefficients of the combination. Using row operations to reduce M to an upper triangular matrix we find that the determinant of $M = (-1)^l(l+1)$. Also since each row contains at most 3 nonzero elements ≤ 2 in absolute value, we see that each $l \times l$ submatrix has determinant $\leq 3^l 2^l$. Hence each element in M^{-1} is less than or equal in absolute value to $6^l/(l+1)$. From our previous discussion, we find that $[p_1^k p_2^r]$, $k+r = l$, can be expressed as a linear combination of the above operators with each coefficient in absolute value $\leq 6^l/l$. Hence Lemma 4.1 is true. □

Next let

$$p_5(x,t,y,s) = q(x,s) - q(x,t),$$
$$p_6(x,t,y,s) = q(y,s) - q(y,t).$$

For fixed N, ϵ, n, η, and $i + l$ an odd positive integer, we prove

Lemma 4.2. *There exists $c_5 > 0$ such that for $g \in C_0^\infty(R^2)$ and $l \geq 1$,*

$$\max\{||[F]g||_2, ||[F]^*g||_2\} \leq A(l)\delta_i^{-1}\beta_1(c_5)||g||_2, \tag{4.1}$$

while if $g \in L^\infty(R^2) \cap L^2(R^2)$ or $g = 1$, then for $l \geq 1$

$$\max\{||[F]g||_{*,\alpha}, ||[F]||^*g||_{*,\alpha}\} \leq A(l)\delta_i^{-1}\beta_1(c_5)||g||_\infty, \tag{4.2}$$

whenever F is one of the representatives:

(a) $$p_2^{(l-1)}p_5,\ p_2^{(l-1)}p_6,\ p_1^{(l-1)}p_5,\ p_1^{(l-1)}p_6$$

$$p_4p_2^{(l-1)}p_5,\ p_4p_2^{(l-1)}p_6,\ p_4p_1^{(l-1)}p_5,\ p_4p_1^{(l-1)}p_6\,.$$

If $l \geq 2$, then

$$\max\{|||[F]g||_2,\ ||[F]^*g||_2\} \leq A(l)\delta_i^{-1}\beta_1(c_5)||g||_2, \tag{4.3}$$

while if $g \in L^\infty(R^2) \cap L^2(R^2)$ or $g = 1$, then for $l \geq 2$

$$\max\{|||[F]g||_{*,\alpha},\ ||[F]^*g||_{*,\alpha}\} \leq A(l-1)\delta_i^{-1}\beta_1(c_5)||g||_\infty, \tag{4.4}$$

whenever F is one of the representatives.

(b) $$p_3p_2^{(l-2)}p_5,\ p_3p_2^{(l-2)}p_6,\ p_3p_1^{(l-2)}p_5,\ p_3p_1^{(l-2)}p_6,$$

$$p_3p_4p_2^{(l-2)}p_5,\ p_3p_4p_2^{(l-2)}p_6,\ p_3p_4p_1^{(l-2)}p_5,\ p_3p_4p_1^{(l-2)}p_6.$$

Proof. Let

$$[F]g(x,t) = \iint_{R^2} \Psi_2(x,t,y,s)g(y,s)dyds,\ (x,t) \in R^2,$$

when F is one of the representatives in (a) or (b) of Lemma 4.2. For fixed $x \in R$ and $H \in L^\infty(R)$ define $\langle F\rangle = \langle F\rangle_{x,H}$, an operator on $C_0^\infty(R)$ by

$$\langle F\rangle k(t) = \int_R \left(\int_R \tilde{\Psi}_2(x,t,y,s)H(y)dy\right) k(s)ds,\ t \in R,$$

when $k \in C_0^\infty(R)$. Here, either $\tilde{\Psi}_2 = \Psi_2$ or $\tilde{\Psi}_2(x,t,y,s) = \Psi_2(y,s,x,t)$. As in section 3, let $\langle F\rangle^*$ be the adjoint of $\langle F\rangle$. The proof of Lemma 4.2 will be based on induction and the following lemma.

Lemma 4.3. *Suppose that either $l = 1$ and F is one of the representatives*

(c) $$p_5,\ p_4p_5,$$

or $l = 2$ and F is one of the representatives

(d) $$p_3p_5,\ p_3p_4p_5.$$

Then

$$\max\{||\langle F\rangle k||_2,\ ||\langle F\rangle^*k||_2\} \leq \delta_i^{-1}A(1)\beta_1(c)||H||_\infty||k||_2, \tag{4.5}$$

when $k \in C_0^\infty(R)$ and

$$\max\{||\langle F\rangle k||_*,\ ||\langle F\rangle^*k||_*\} \leq \delta_i^{-1}A(1)\beta_1(c)||H||_\infty||k||_\infty, \tag{4.6}$$

for $k \in L^\infty(R) \cap L^2(R)$ or $k = 1$.

Proof. As in previous lemmas, we shall use the T1 theorem to prove Lemma 4.3. We prove one dimensional versions of (1.9)-(1.11) only for $F = p_5$, $\tilde{\Psi} = \Psi$, since the other proofs are essentially the same. As in section 3 we note that $\langle p_5\rangle 1$, $\langle p_5\rangle^* 1$, are well defined, because all integrals are absolutely convergent. To show $\langle p_5\rangle 1 \in \mathrm{BMO}(R)$ we first show for fixed $x \in R$ that the measure ν, defined on $R \times (0,\infty)$ by

$$d\nu(t,\lambda) = \lambda^{-1}\langle(\phi')^t_\lambda,\ \langle p_5\rangle 1\rangle^2 dtd\lambda,$$

is a Carleson measure on $R \times (0,\infty)$. In fact we prove

$$\nu(G) \leq \delta_i^{-2}\beta_1(i,1,c)^2 a_2^2||H||_\infty^2 r, \tag{4.7}$$

whenever

$$G = \{t : |t - t_0| < r\} \times (0, r) = J \times (0, r),$$

$\phi' = \frac{d\phi}{dt}$, and $(\phi)^t_\lambda$, is defined as in section 3. To prove (4.7), let $\tilde{J} = \{t : |t - t_0| < 4r\}$, and set

$$P(x, t, \lambda) = \int_{\mathbf{R}} (\phi')^t_\lambda(\rho)\, q(x, \rho) d\rho,$$

for fixed $x \in R$. Then, since all integrals are absolutely convergent,

(4.8)

$$\langle(\phi')^t_\lambda, \langle p_5\rangle 1\rangle = \int_{\mathbf{R}} |u|^{-\bar{\sigma}} \left\{\int_{\mathbf{R}} (\phi')^t_\lambda(\rho)(q(x, \rho + u) - q(x, \rho))d\rho\right\} \cdot \left(\int_{\mathbf{R}} \Omega_i(x, y, u) H(y) dy\right) du$$

$$= \int_{\mathbf{R}} |u|^{-\bar{\sigma}} \{P(x, t + u, \lambda) - P(x, t, \lambda)\} \left(\int_{\mathbf{R}} \Omega_i(x, y, u) H(y) dy\right) du$$

$$= \int_{\{|u|<\lambda\}} \cdots + \int_{\{\lambda<|u|<r\}} \cdots + \int_{\{r<|u|\}} \cdots = (K_1 + K_2 + K_3)(x, t, \lambda).$$

Recall that $\bar{\sigma} = \bar{\sigma}(i, 1)$ in (4.8) . We note that

(4.9)
$$\begin{aligned} |P(x, \tau, \lambda)| &= \left|\int_{\mathbf{R}} (\phi')^\tau_\lambda(\rho)(q(x, \rho) - q(x, \tau))d\rho\right| \\ &\le c\lambda^{-1} \int_{\{|\rho-\tau|<\lambda\}} |q(x, \rho) - q(x, \tau)| d\rho = c\lambda^{-1} W(x, \tau, \lambda). \end{aligned}$$

Similarly, for $|u| \le \lambda$

(4.10)
$$\begin{aligned} &|P(x, \tau + u, \lambda) - P(x, \tau, \lambda)| \\ &\le \left|\int_{\mathbf{R}} \{(\phi')^{\tau+u}_\lambda - (\phi')^\tau_\lambda\}(\rho)[q(x, \rho) - q(x, \tau)] d\rho\right| \\ &\le c|u|\lambda^{-2} W(x, \tau, 2\lambda). \end{aligned}$$

In view of (4.10), (3.17), (2.3) with $\lambda = -1, \nu = 1 + 2\alpha$, and the fact that $\sigma - \bar{\sigma} = 2\alpha$, we deduce

(4.11)
$$\begin{aligned} |K_1(x, t, \lambda)| &\le \tfrac{c^{i+1} W(x, t, 2\lambda)}{\lambda^{2\delta_i}} \|H\|_\infty \int_0^\lambda u^{1+2\alpha} \left(\int_{\mathbf{R}} V_{i,1}(z, u)|z|^{-1} dz\right) du \\ &\le \|H\|_\infty \tfrac{\beta_1(c)}{\lambda^{2\delta_i}} W(x, t, 2\lambda) \left(\int_0^\lambda u^{-\alpha} du\right) \\ &\le \|H\|_\infty \lambda^{-(1+\alpha)} \delta_i^{-1} \beta_1(c) W(x, t, 2\lambda). \end{aligned}$$

From (4.11), Schwarz's inequality, and (2.24) it follows that

(4.12)
$$\iint_G K_1(x,t,\lambda)^2\lambda^{-1}dtd\lambda \le ||H||_\infty^2 \delta_i^{-2}\beta_1^2(c) \int_J \int_0^r \lambda^{-(3+2\alpha)}W(x,t,\lambda)^2 dtd\lambda$$

$$\le ||H||_\infty^2 \delta_i^{-2}\beta_1^2(c) \int_J \int_0^r \left(\int_{-\lambda}^{\lambda} |q(x,t+\tau)-q(x,t)|^2 d\tau\right) \lambda^{-(2+2\alpha)}dtd\lambda$$

$$\le ||H||_\infty^2 \delta_i^{-2}\beta_1^2(c) \int_J \int_{-r}^{r} |q(x,t+\tau)-q(x,t)|^2|\tau|^{-(1+2\alpha)}d\tau dt$$

$$\le ||H||_\infty^2\, a_2^2\delta_i^{-2}\beta_1^2(c) r,$$

for c large enough, where the last integral was obtained by interchanging the order of integration.

Let Mg denote the Hardy-Littlewood maximal function of $g \in L^1(R)$. Then from (4.9), (3.17), and (2.3) with $\lambda = -1, \nu = 2\alpha$, we get for $(t,\lambda) \in G$

(4.13)
$$|K_2(x,t,\lambda)|$$

$$\le \frac{||H||_\infty c^{i+1}}{\lambda\delta_i} \int_{\{\lambda<|u|<r\}} (W(x,t+u,\lambda)+W(x,t,\lambda))$$

$$\cdot\left(\int_{\boldsymbol{R}} V_{i,1}(z,u)|z|^{-1}dz\right)|u|^{2\alpha}du$$

$$\le ||H||_\infty\, \beta_1(c)\lambda^{-1}\delta_i^{-1} \int_{\{\lambda<|u|<r\}} (W(x,t+u,\lambda)+W(x,t,\lambda))|u|^{-(1+\alpha)}du.$$

If $T_k = \{u : \lambda 2^k < |u| \le 2^{k+1}\lambda\}$, $k = 0,1,2,\ldots,j$, where $\lambda 2^{j-1} \le r < \lambda 2^j$, we estimate the above integral as follows:

$$\int_\lambda^r (W(x,t+u,\lambda)+W(x,t-u,\lambda))|u|^{-(1+\alpha)}du$$

$$\le \sum_{k=0}^{j} \int_{T_k} (W(x,t+u,\lambda)+W(x,t-u,\lambda))|u|^{-(1+\alpha)}du$$

$$\le c\sum_{k=0}^{j}(2^k\lambda)^{-\alpha}M[\chi_J W(x,\cdot,\lambda)](t) \le c\lambda^{-\alpha}M[\chi_J\, W(x,\cdot,\lambda)](t)\,.$$

Putting this inequality in (4.13) we obtain

$$|K_2(x,t,\lambda)| \le ||H||_\infty\, \beta_1(c)\lambda^{-(1+\alpha)}\delta_i^{-1}M[\chi_J\, W(x,\cdot,\lambda)](t). \tag{4.14}$$

From the Hardy-Littlewood maximal theorem and (4.14) we find as in (4.12),

$$\iint_G K_2(x,t,\lambda)^2\lambda^{-1}dtd\lambda$$

$$\text{(4.15)} \qquad \le \|H\|_\infty^2\, \beta_1^2(c)\delta_i^{-2} \int_0^r \left(\int_{\mathbf{R}} M[\chi_{\bar{J}} W(x,\cdot,\lambda)]^2(t)dt\right)\lambda^{-(3+2\alpha)}d\lambda$$

$$\le \|H\|_\infty^2\, \beta_1^2(c)\delta_i^{-2} \int_0^r \left(\int_{\bar{J}} W(x,t,\lambda)^2dt\right)\lambda^{-(3+2\alpha)}d\lambda$$

$$\le a_2^2\delta_i^{-2}\beta_1^2(c)\|H\|_\infty^2\, r.$$

To estimate K_3 note from (0.4) that

$$W(x,\tau,\lambda) \le ca_2\lambda^{1+\alpha}, \quad \tau \in R,$$

so as in (4.13)

$$|K_3(x,t,\lambda)|$$

$$\le \|H\|_\infty\, \beta_1(c)\, \lambda^{-1}\delta_i^{-1}\int_{\{r<|u|\}} (W(x,t+u,\lambda)+W(x,t,\lambda))|u|^{-(1+\alpha)}du$$

$$\le \|H\|_\infty\, \beta_1(c)\lambda^\alpha\delta_i^{-1}\, a_2\, r^{-\alpha}.$$

Hence,

$$\text{(4.16)} \qquad \iint_G K_3(x,t,\lambda)^2\,\lambda^{-1}dtd\lambda \le a_2^2\,\delta_i^{-2}\,\beta_1^2(c)\|H\|_\infty^2\, r.$$

From (4.8), (4.12), (4.15), and (4.16), we deduce that (4.7) is valid. From (4.7) and an argument of C. Fefferman (see [T, ch 12, section 3]) we conclude that (4.6) holds when $F = p_5$ and $k = 1$. From a similar proof we get (4.6) when F is one of the other representatives in (c), (d), and $k = 1$. (4.6) for $k = 1$ and $\langle F\rangle^*$, when F is as in (c), (d), iollows from the observation that $\langle F\rangle^* = \pm\langle F\rangle$.

Next given $\lambda > 0$, let

$$L(t) = L(t,\lambda)$$

$$= \int_{\{|u|<8\lambda\}} |u|^{-\bar{\sigma}}(q(x,t+u)-q(x,t))\left(\int_{\mathbf{R}} \Omega_i(x,y,u)H(y)dy\right)du,$$

when $t \in R$. Then for $|s_0 - t_0| \le 4\lambda$,

$$|\langle(\phi')_\lambda^{t_0},\ \langle p_5\rangle(\phi_\lambda^{s_0})\rangle|$$

$$\text{(4.17)} \qquad \le |\langle(\phi')_\lambda^{t_0},\ \langle p_5\rangle(\phi_\lambda^{s_0}) - \phi_\lambda^{s_0}L\rangle| + |\langle(\phi')_\lambda^{t_0},\ \phi_\lambda^{s_0}L\rangle|$$

$$= T_1 + T_2\,.$$

Using (0.4) and arguing as in the estimate of K_1, we get for $|s_0 - t_0| \le 4\lambda$,

$$|\langle p_5\rangle(\phi_\lambda^{s_0}) - \phi_\lambda^{s_0} L|(t)$$

$$\le \int_{\{|u|<8\lambda\}} |u|^{-\sigma} |q(x,t+u) - q(x,t)|\; |\phi_\lambda^{s_0}(t) - \phi_\lambda^{s_0}(t+u)|$$

$$\cdot \left| \int_{\mathbf{R}} \Omega_i(x,y,u) H(y)dy \right| du$$

$$\le a_2 ||H||_\infty c^{i+1} \lambda^{\alpha-2} \delta_i^{-1} \int_0^{8\lambda} u^{1+2\alpha} \left(\int_{\mathbf{R}} V_{i,1}(z,u)|z|^{-1} dz \right) du$$

$$\le a_2 ||H||_\infty\, \beta_1(c) \lambda^{\alpha-2} \delta_i^{-1} \left(\int_0^{8\lambda} u^{-\alpha} du \right) \le a_2 ||H||_\infty\, \beta_1(c) \delta_i^{-1} \lambda^{-1} .$$

Hence

$$|T_1| \le a_2 ||H||_\infty\, \beta_1(c) \delta_i^{-1} \lambda^{-1} . \tag{4.18}$$

To estimate T_2 with x fixed and $|s_0 - t_0| < 4\lambda$, let

$$\xi(\tau) = (\phi')^{t_0}(-\tau)\phi^{s_0}(-\tau), \quad \tau \in R,$$

$$P_1(x,t) = \langle \xi_\lambda^t, q(x,\cdot) - q(x,t_0)\rangle .$$

Clearly,

$$\left| \frac{d}{dt} P_1(x,t) \right| \le c a_2 \lambda^{\alpha-2} .$$

Using this inequality, interchanging the order of integration as in (4.8), and arguing as (4.11), we obtain

$$|T_2| = | \int_0^{8\lambda} |u|^{-\sigma} (P_1(x,u) - P_1(x,0)) \left(\int_{\mathbf{R}} \Omega_i(x,y,u) H(y) dy \right) du| \tag{4.19}$$

$$\le c a_2 \lambda^{\alpha-2} c^{i+1} \delta_i^{-1} ||H||_\infty \left\{ \int_{-8\lambda}^{8\lambda} u^{1+2\alpha} \left(\int_{\mathbf{R}} V_{i,1}(z,u)|z|^{-1} dz \right) du \right\}$$

$$\le a_2 \lambda^{\alpha-2} \beta_1(c) ||H||_\infty\, \delta_i^{-1} \left(\int_0^{8\lambda} u^{-\alpha} du \right)$$

$$\le a_2 \beta_1(c) ||H||_\infty \delta_i^{-1} \lambda^{-1} .$$

Putting (4.18), (4.19), into (4.17) we deduce that

$$|\langle (\phi')_\lambda^{t_0},\ \langle p_5\rangle(\phi_\lambda^{s_0})\rangle| \le a_2 ||H||_\infty\, \beta_1(c) \delta_i^{-1} \lambda^{-1} , \tag{4.20}$$

for fixed $x \in R$ and $|s_0 - t_0| \le 4\lambda$.

If $|s_0 - t_0| > 4\lambda$, then from (0.4), (3.31), and Lemma 2.2 we get for $|s - s_0| < \lambda$, $|t - t_0| < \lambda$,

$$\int_{\mathbf{R}} |\tilde{\Psi}_2(x,t,y,s) - \tilde{\Psi}_2(x,t_0,y,s_0)|dy \le \beta_1(c)\delta_i^{-1}\, a_2\lambda^\alpha |s_0 - t_0|^{-(1+\alpha)}.$$

Hence, if $e = \int_{\mathbf{R}} \tilde{\Psi}_2(x,t_0,y,s_0)H(y)dy$, then

$$\begin{aligned} &(4.21)\\ &|\langle(\phi')_\lambda^{t_0},\ \langle p_5\rangle(\phi_\lambda^{s_0})\rangle| = |\langle(\phi')_\lambda^{t_0},\ \langle p_5\rangle(\phi_\lambda^{s_0}) - e\rangle| \\ &\le \|H\|_\infty \iint_{\mathbf{R}^2} |(\phi')_\lambda^{t_0}(t)\phi_\lambda^{s_0}(s)| \left(\int_{\mathbf{R}} |\tilde{\Psi}_2(x,t,y,s) - \tilde{\Psi}_2(x,t_0,y,s_0)|dy\right) dsdt \\ &\le \beta_1(c)\delta_i^{-1}\|H\|_\infty\, a_2\lambda^\alpha |s_0 - t_0|^{-(1+\alpha)}, \end{aligned}$$

where we have used in addition to the above inequality the fact that

$$\int_{\mathbf{R}} (\phi')_\lambda^{t_0} dt = 0, \quad \int_{\mathbf{R}} \phi_\lambda^{s_0} dt = 1.$$

From (4.20), (4.21), we get one dimensional analogues of (1.9) when $F = p_5$. The proof of (1.9) when either $l = 1$ and $F = p_4p_5$, or $l = 2$, F as in (d), is essentially the same. Since $\langle F\rangle^* = \pm\langle F\rangle$, we conclude that (1.9), (1.10), hold when $T = \langle F\rangle$, and F is as in (c), (d). Since we have already shown that $\langle F\rangle 1, \langle F\rangle^* 1 \in \mathrm{BMO}(R)$, we conclude that (4.5) is true. (4.6) follows from (4.5) in a now routine way, by applying Peetre's argument and using the previous estimates on our kernels. The proof of Lemma 4.3 is complete. □

Next we prove Lemma 4.2. To do so we allow x to vary and consider $\langle F\rangle k$ as a function on R^2, when $k \in L^\infty(R) \cap L^2(R)$ or $k = 1$. We claim that

$$\max\{\|\langle F\rangle k\|_{*,\alpha},\ \|\langle F\rangle^* k\|_{*,\alpha}\} \le A(1)\delta_i^{-1}\beta_1(c)\|H\|_\infty \tag{4.22}$$

when $l = 1$ and F is one of the representatives in (c) or $l = 2$ and F is as in (d). The proof is similar to the proof of (3.53). In (4.22), $\langle F\rangle^*$ now denotes the adjoint of $\langle F\rangle$, considered as an operator on $L^2(R^2)$. Let $Q \subseteq \tilde{Q} \subseteq Q^\sharp$, be as in section 2 with $\tilde{Q} = I_1 \times I_2$. Then from Peetre's argument, (4.5), and our kernel estimates we find for $k_2 = k\,\chi_{(\mathbf{R}\setminus I_2)}$,

$$\int_{\{|t-t_0|<d/2\}} |\langle F\rangle k(x,t) - \langle F\rangle k_2(x,t_0)|dt \le A(1)\delta_i^{-1}\beta_1(c)\|H\|_\infty,$$

and

$$|\langle F\rangle k_2(x,t_0) - \langle F\rangle k_2(x_0,t_0)| \le \delta_i^{-1}\beta_1(c)A(1)\|H\|_\infty,$$

when $(x,t_0) \in Q$. Integrating with respect to x we obtain from these inequalities that

$$\iint_Q |\langle F\rangle k(x,t) - \langle F\rangle k_2(x_0,t_0)|dtdx \le d^{1+\alpha}\delta_i^{-1}\beta_1(c)A(1)\|H\|_\infty,$$

which implies (4.22) for $\langle F\rangle$. (4.22) for $\langle F\rangle^*$ follows from (4.22) for $\langle F\rangle$ and the definition of $\tilde{\Psi}$. . From (4.22) with $H = 1$, we have

$$\max\{|||[F]1||_{*,\alpha}\,,\ ||[F]^*||_{*,\alpha}\} \le A(1)\delta_i^{-1}\beta_1(c), \tag{4.23}$$

when $l = 1$, F as in (c), or $l = 2$, F as in (d). For fixed $(y_0, s_0) \in Q^\sharp$, recall that $\theta_\lambda^{(y_0,s_0)} = c\phi_{\lambda\alpha}^{y_0}\phi_\lambda^{s_0}$, so if $H = c\phi_{\lambda\alpha}^{y_0}$, $k = \phi_\lambda^{s_0}$, then

$$[F](\theta_\lambda^{(y_0,s_0)}) = \langle F\rangle_H(k) \text{ on } R^2.$$

It follows from this equality for $(y_0, s_0) \in Q^\sharp$ and k_2 as above that for $\lambda = d/2$,

$$\begin{aligned} &|\langle (D\theta)_\lambda^{(x_0,t_0)},\ [F](\theta_\lambda^{(y_0,s_0)})\rangle| \\ &= |\langle (D\theta)_\lambda^{(x_0,t_0)},\ [F]_H(k) - [F]_H(k_2)(x_0,t_0)\rangle| \\ &\le c\lambda^{-(1+\alpha)} \iint\limits_Q |\langle F\rangle_H(k) - \langle F\rangle_H(k_2)(x_0,t_0)|dxdt \\ &\le A(1)\delta_i^{-1}\beta_1(c)\lambda^{-(1+\alpha)}. \end{aligned} \tag{4.24}$$

(4.24) also holds with $[F]$ replaced by $[F]^*$, as we see from the definition of $\tilde{\Psi}_2$. If $(y_0, s_0) \in R^2 \setminus Q^\sharp$ we see as in (2.35) that for some j, $1 \le j \le 3$, we have supp $(\theta_\lambda^{(y_0,s_0)}) \subseteq E_j$. Arguing as in (3.29) and (3.34) we deduce

$$|\langle (D\theta)_\lambda^{(x_0,t_0)}[F](\theta_\lambda^{(y_0,s_0)})\rangle| \le c^{i+2}A(1)\delta_i^{-1}\sum_{j=1}^{3}(\mathcal{H}_j)_\lambda(x_0 - y_0, s_0 - t_0). \tag{4.25}$$

where $\mathcal{H}_j$ are defined as in section 2 relative to $i, l = 1$. By the same argument we also get (4.25) with $[F]$ replaced by $[F]^*$. From (4.23)-(4.25) and the T1 theorem we conclude that (4.1) is valid when $l = 1$ (F as in (c)) and (4.3) is valid when $l = 2$ (F as in (d)). (4.2), (4.4) follow from (4.1), (4.3), by Peetre's argument.

To continue the proof of Lemma 4.2, we note that $p_6 - p_5 = p_1 - p_2$. Since we have already shown that p_1, p_2, and p_5 are the representatives of bounded operators, it follows that p_6 also is the representative of a bounded operator. From this note we conclude that Lemma 4.2 is valid for $l = 1$, F as in (a), and $l = 2$, F as in (b). We proceed by induction. Suppose that Lemma 4.2 is true for pairs (i, l) whenever $1 \le l \le m$, and $l + i$ is an odd positive integer. If $l = m + 1$ we first suppose that F is one of the representatives:

(a') $p_1^m p_5$, $p_4 p_1^m p_5$,

or

(b') $p_3 p_1^{m-1} p_5$, $p_4 p_3 p_1^{m-1} p_5$

when $m \ge 2$. Integrating by parts as in (3.47) we obtain for $m \ge 1$

$$[p_1^m p_5]_{i,\eta}1 = m[p_1^{(m-1)}p_5]_{i+1,\bar{\eta}}(q'),$$

$$[p_4 p_1^m p_5]_{i,\eta} 1 = m[p_4 p_1^{(m-1)} p_5]_{i+1,\bar\eta}(q'),$$

and for $m \geq 2$

$$[p_3 p_1^{(m-1)} p_5]_{i,\eta} 1 = [p_1^{m-1} p_5]_{i+1,\bar\eta} 1 + (m-1)[p_3 p_1^{m-2} p_5]_{i+1,\bar\eta}(q'),$$

$$[p_4 p_3 p_1^{m-1} p_5]_{i,\eta} = [p_4 p_1^{m-1} p_5]_{i+1,\bar\eta} 1 + (m-1)[p_4 p_3 p_1^{m-2} p_5]_{i+1,\bar\eta}(q').$$

Applying the induction hypothesis and (4.2), (4.4) we conclude that

$$||[F]_{i,\eta} 1||_{*,\alpha} \leq m\tilde{A}(m+1)\delta_i^{-1}\beta_1(i+1, m, c_5), \tag{4.26}$$

when F is as in (a') or (b'). Here $\tilde{A}(m+1) = A(m+1)$ when F is as in (a) of Lemma 4.2 and $\tilde{A}(m+1) = A(m)$ for F as in (b) of Lemma 4.2 . Given $\lambda > 0$ let $\hat\eta = (\hat\eta_j)$ where $\hat\eta_{i+1} = (\phi_{\lambda^\alpha}^{y_0}, +)$ and $\hat\eta_j = \eta_j$, $j \neq i+1$. Then

$$[p_1^m p_5]_{i,\eta}(\theta_\lambda^{(y_0,s_0)}) = m[p_1^{(m-1)} p_5]_{i+1,\hat\eta}(c\phi_\lambda^{s_0} q'),$$

with similar equalities holding for the other operators in (a'), (b'). Using the induction hypothesis once again it follows that

$$||[F]_{i,\eta}(\theta_\lambda^{(y_0,s_0)})||_{*,\alpha} \leq cm\delta_i^{-1}\lambda^{-(1+\alpha)}\tilde{A}(m+1)\beta_1(i+1, m, c_5). \tag{4.27}$$

From (4.27) we see as in (4.24) that for $(y_0, s_0) \in Q^\sharp$, $\lambda = d/2$, and F as in (a'), (b'),

$$|\langle (D\theta)_\lambda^{(x_0,t_0)}, [F](\theta_\lambda^{(y_0,s_0)})\rangle| \leq cm\tilde{A}(m+1)\delta_i^{-1}\beta_1(i+1, m, c_5)\lambda^{-(1+\alpha)}. \tag{4.28}$$

As in (4.25) it also follows from our kernel estimates that for $(y_0, s_0) \in R^2 \setminus Q^\sharp$ we have

(4.29)

$$|\langle (D\theta)_\lambda^{(x_0,t_0)}, [F](\theta_\lambda^{(y_0,s_0)})\rangle| \leq c^{i+2}\tilde{A}(m+1)\delta_i^{-1}\sum_{j=1}^{3}(\mathcal{H}_j)_\lambda(x_0 - y_0, s_0 - t_0),$$

where $\mathcal{H}_j$, $1 \leq j \leq 3$, are defined as in section 2 relative to i, $m+1$.

To obtain similar estimates for $[F]^*$ we observe that

$$\begin{aligned} p_1^m - p_2^m &= (p_1 - p_2)(p_1^{m-1} + p_1^{m-2}p_2 + \dots + p_2^{m-1}) \\ &= (p_6 - p_5)(p_1^{m-1} + \dots + p_2^{m-1}). \end{aligned} \tag{4.30}$$

Thus

$$\begin{aligned} [p_1^m p_5]^* &= [p_2^m p_5]^* + [p_5(p_6 - p_5)(p_1^{m-1} + \dots + p_2^{m-1})]^* \\ &= [p_2^m p_6]^* + [p_2^m(p_2 - p_1)]^* + [p_5(p_6 - p_5)(p_1^{m-1} + \dots + p_2^{m-1})]^* \\ &= S_1 + S_2 + S_3. \end{aligned} \tag{4.31}$$

Now from Lemma 4.1 we see for $g \in L^2(R^2) \cap L^\infty(R^2)$, or $g = 1$ that

$$||S_2 g||_{*,\alpha} \leq 2\delta_i^{-1}\tilde{A}(m+1)\beta_1(i, m+1, c_4)||g||_\infty. \tag{4.32}$$

Also, the kernel of S_3 can be estimated using (0.1) for q and (3.17). Doing this and using (2.23b) of Proposition 2.5 we obtain for $g \in L^\infty(R^2) \cap L^2(R^2)$ that

$$(4.33) \qquad \| S_3 g\|_{*,\alpha} \leq m\delta_i^{-1}\tilde{A}(m+1)\beta_1(i, m-1, c_2)\|g\|_\infty.$$

Now for $\tilde{\eta}$ as in (3.46) we note that

$$S_1 1 = [p_2^m p_6]^*_{i,\eta} 1 = m(-1)^{m+1}[p_1^{m-1}p_5]_{i+1,\tilde{\eta}}(q').$$

From this equality and the induction hypothesis we find

$$(4.34) \qquad \|S_1 1\|_{*,\alpha} \leq m\delta_i^{-1}\tilde{A}(m+1)\beta_1(i+1, m, c_5).$$

Next define $\eta' = (\eta'_j)$ by, $\eta'_{i+1} = (\phi^{y_0}_{\lambda\alpha}, -)$ and $\eta'_j = \eta_j, j \neq i+1$. Then

$$S_1(\theta_\lambda^{(y_0,s_0)}) = m(-1)^{m+1}[p_1^{m-1}p_5]_{i+1,\eta'}(c\phi_\lambda^{s_0} q'),$$

so again using the induction hypothesis we get

$$(4.35) \qquad \| S_1(\theta_\lambda^{(y_0,s_0)})\|_{*,\alpha} \leq cm\delta_i^{-1}\lambda^{-(1+\alpha)}\tilde{A}(m+1)\beta_1(i+1, m, c_5)$$

when $(y_0, s_0) \in Q^\sharp$. From (4.30)-(4.35) we deduce that (4.26),(4.28), are valid with $[F]$ replaced by $[F]^*$ and β_1 by $\tilde{\beta}_1$ where $F = p_1^m p_5$ and

$$\tilde{\beta}_1(i+1, m, c_5) \equiv cm\beta_1(i+1, m, c_5) + 2\beta_1(i, m+1, c_4) + m\beta_1(i, m-1, c_2).$$

By a similar argument we get (4.26),(4.28), for $[F]^*$ with β_1 replaced by $\tilde{\beta}_1$ when F is one of the other representatives in (a'), (b'). Thus (4.28) remains true with $[F]$ replaced by $[F]^*$, and β_1 by $\tilde{\beta}_1$, when F is as in (a'), (b'). Using our kernel estimates it also can be checked that (4.29) holds for $[F]^*$ and F as in (a'), (b'). In view of (4.26), (4.28),(4.29), we can once again apply the T1 Theorem. Doing this we get for F as in (a'), (b'), and $g \in C_0^\infty(R^2)$ that

$$(4.36) \qquad \|[F]_{i,\eta} g\|_2 \leq c\delta_i^{-1}\tilde{A}(m+1)\tilde{\beta}_1(i+1, m, c_5)\|g\|_2$$

$$\leq (c_5)^{-1/2}\, \delta_i^{-1}\tilde{A}(m+1)\beta_1(i, m+1, c_5)\|g\|_2 ,$$

for c_5 large enough, since

$$(m+1)c_5\beta_1(i+1, m, c_5) \leq \beta_1(i, m+1, c_5).$$

Clearly (4.36) implies (4.1), (4.3), for $(i, m+1)$ and F as in (a'), (b').

The other representatives can be handled by arguing as in (4.30)-(4.31). That is, any other representative in (a) or (b) can be expressed as a representative in (a') or (b') plus the representatives of bounded operators whose norms can be estimated using Lemma 4.1 or Proposition 2.5. From (4.36) and these estimates it follows for c_5 large enough and $g \in C_0^\infty(R^2)$ that

$$(4.37) \qquad \|[F]_{i,\eta} g\|_2 \leq (c_5)^{-1/4}\, \delta_i^{-1}\tilde{A}(m+1)\beta(i, m+1, c_5)\|g\|_2 ,$$

whenever F is as in (a) or (b). Hence (4.1), (4.3) are valid for $(i, m+1)$ provided c_5 is sufficiently large. (4.2), (4.4) follow from (4.37) and Peetre's argument. With $c_5 > 0$ now fixed, we conclude by induction that Lemma 4.2 is true. □

5. Proof of Theorem 1.

We prove Theorem 1 only for Λ_1^m since the proofs for Λ_j^m, $2 \le j \le 2n$, are essentially identical. For the moment we continue to work in R^2 and shall use the same notation as in sections 2-4. We first prove Theorem 1 with Λ_1^m replaced by $[(p_1+p_5)^{2m+1}]$ when $l = 2m+1$, $i = 0$, and $\eta = (\eta_k)$, $\eta_k = (1,+)$, $k = 1,2,\dots$. To do this observe from the mean value theorem of calculus for higher derivatives and (0.1), (0.4), that for $m = 1,2,\dots$, we have at (x,t,y,s)

$$(5.1)\quad |(p_1+p_5)^{2m+1} - p_1^{2m+1} - (2m+1)p_1^{2m}p_5| \le cm^2(|p_1|+|p_5|)^{2m-1}|p_5|^2$$

$$\le cm^2 4^{2m}(a_1^{2m-1}|x-y|^{2m-1} + a_2^{2m-1}|s-t|^{(2m-1)\alpha})|p_5|^2 = p_7.$$

If $m = 0$, put $p_7 \equiv 0$ and note that (5.1) remains true. We now put $l = 2m+1$, $i = 0$, in Lemmas 4.1 and 4.2. With ϵ, n, N, still fixed we see for $g \in C_0^\infty(R^2)$ that

$$\max\{||[p_1^{2m+1}]g||_2, ||[p_1^{2m}p_5]g||_2\} \le A(2m+1)\beta_1(0,2m+1,c)||g||_2$$

for c large enough. If $m \ge 1$, then from Proposition 2.5 with $l = 2m-1$, $i = 0$, and $l = 0$, $i = 0$, we also find that

$$||[p_7]_{0,2m+1}(g)||_2 \le (a_1^{2m-1}\beta_1(0,2m+1,c) + c^m a_2^{2m-1})(a_1^2+a_2^2)||g||_2.$$

From these inequalities and (5.1) we conclude for $m \ge 1$ and $g \in C_0^\infty(R^2)$ that

(5.2)
$$||[(p_1+p_5)^{2m+1}]g||_2 \le \{A(2m+1)\beta_1(0,2m+1,c) + c^m a_2^{2m-1}(a_1^2+a_2^2)\}||g||_2,$$

for c large enough. If $m = 0$, the above inequality remains true provided the second term on the right-hand side of (5.2) is replaced by zero. Extend $[(p_1+p_5)^{2m+1}]$ to a bounded operator on $L^2(R^2)$ in the usual way. Then (5.2) is true for $g \in L^2(R^2)$.

Next we allow ϵ and N to vary and keep $l = 2m+1$, $i = 0$. We claim that for $g \in L^2(R^2)$ or $g = 1$,

$$(5.3)\qquad \lim_{\substack{\epsilon\to 0\\ N\to\infty}} [(p_1+p_5)^{2m+1}]g, \text{ and for fixed } N, \lim_{\epsilon\to 0}[(p_1+p_5)^{2m+1}]g,$$

exist in the norm of $L^2(R^2)$. We prove only the first claim. If

$$(5.4)\qquad q(x,\cdot) = I_\alpha \star \bar{b}(x,\cdot),$$

where I_α is as in section 1, then because $q(x,\cdot)$ has compact support we have

$$(5.5)\qquad |\bar{b}(x,t)| = \left| c\int_{-\infty}^{\infty} \frac{q(x,s)-q(x,t)}{|s-t|^{1+\alpha}}ds \right| = O(|t|^{-(1+\alpha)}) \text{ as } t \to +\infty.$$

Since $\bar{b}(x,\cdot) \in \text{BMO}\,(R)$ (see (0.3)), it follows from (5.5) that $\bar{b}(x,\cdot) \in L^r(R)$ for almost every $x \in R$ and $1 \le r < \infty$. The first equality in (5.5) can be proved using (5.4) and Fourier transforms. Indeed, observe for $l = 1, i = 0,2,\dots$, $j = \sqrt{-1}$, and ϵ, N, i, x, fixed that

$$(5.6)\quad \widehat{\langle p_5\rangle 1}(x,\tau) = \int_R\int_R |u|^{-[(n+i)\alpha+1]}\Omega_i(x,y,u)(e^{ju\tau}-1)\hat{q}(x,\cdot)(\tau)dudy$$

$$= c\left(\int_{R}\int_{R}|u|^{-[(n+i)\alpha+1]}\Omega_i(x,y,u)(e^{ju\tau}-1)|\tau|^{-\alpha}dudy\right)\hat{\hat{b}}(x,\cdot)(\tau)$$

$$= H(x,\tau,\epsilon,N)\hat{\hat{b}}(x,\cdot)(\tau).$$

Here we have used (5.4). if $\eta = (\eta_k)$, $\eta_k = (1,+)$, $k = 1,2,...$, then clearly

$$H(x,\tau,\epsilon,N) \le c\int_{R}^{\infty}\int_{R}^{\infty}\int_{\xi_i}^{\infty}\cdots\int_{\xi_2}^{\infty}|\xi_1|^{n-2}|u|^{-[(n+i)\alpha+1]}\frac{|e^{j\tau u}-1|}{|\tau|^{\alpha}}$$

$$\cdot\exp\left(\frac{-|\xi_1|^{1/\alpha}}{|u|}\right)d\xi_1\cdots d\xi_i dudy$$

$$= c(i,\alpha,n),$$

as follows easily from changing variables in the above integrals. From this fact we deduce for $x,\tau \in R$, that $\lim_{\substack{\epsilon\to 0\\ N\to\infty}} H(x,\tau,\epsilon,N) = c$. Using this equality, dominated convergence, (5.6), and Plancherel's theorem, we conclude that

$$\lim_{\substack{\epsilon\to 0\\ N\to\infty}} [p_5]_{\epsilon,N}1(x,t) = c\bar{b}(x,t),$$

in the norm of $L^2(R^2)$ when $l = 1$, $i = 0,2,4,...$

Now from the above equality, Lemma 4.2, and the previous estimates on our kernels it follows in a standard way that $\lim_{\substack{\epsilon\to 0\\ N\to\infty}} [p_5]_{\epsilon,N}(\chi_G)$ exists in the norm of $L^2(R^2)$, for each square G. Because linear combinations of these characteristic functions are dense in $L^2(R^2)$ we conclude from Lemma 4.2 that $\lim_{\substack{\epsilon\to 0\\ N\to\infty}} [p_5]g$ exists in the norm of $L^2(R)$. We now proceed by induction. Suppose for $\eta = \{(1,+)\}$ and $l \le k$ that

$$\lim_{\substack{\epsilon\to 0\\ N\to\infty}} [p_1^{l-1}p_5]_{i,l,\epsilon,N}(g) \text{ exists,} \tag{5.7}$$

in the norm of $L^2(R^2)$ whenever $i+l$ is an odd positive integer and either $g \in L^2(R^2)$ or $g = 1$. Then since for fixed ϵ, N,

$$[p_1^k p_5]_{i,k+1}(1) = k[p_1^{k-1}p_5]_{i+1,k}(q'),$$

it follows by induction that (5.7) holds whenever $g = 1$, $\eta = \{(1,+)\}$, and $l = k+1$. Using our kernel estimates again and Lemma 4.2, as in the case $l = 1$, we then obtain (5.7) for $g \in L^2(R^2)$. and $l = k+1$. Thus by induction the limit in (5.7) exists. Next we consider $[p_1^l]$. Clearly, $[p_1^l]1 = 0$, $l = 0$, $i = 1,3,...$, since $\eta = \{(1,+)\}$ and ϵ, N, are fixed. This fact, Lemma 4.1, and our kernel estimates imply for $g \in L^2(R^2)$ that $\lim_{\substack{\epsilon\to 0\\ N\to\infty}} [p_1^l]g$ exists in the norm of $L^2(R^2)$ when $l = 0$, $i = 1,3,....$ Using induction and arguing as above we get for $g \in L^2(R^2)$ or $g = 1$, $l+i$ odd, that

$$\lim_{\substack{\epsilon\to 0\\ N\to\infty}} [p_1^l]\, g \text{ exists,} \tag{5.8}$$

in the norm of $L^2(R^2)$. Finally, using Proposition 2.5 and dominated convergence, we easily get for $i = 0,\ l = 2m+1$, and η as above, that

$$\lim_{\substack{\epsilon\to 0\\ N\to\infty}} [p_7]g \text{ exists,} \tag{5.9}$$

when $g \in L^2(R^2)$ or $g = 1$. From (5.7)-(5.9) with $l = 2m+1$ and (5.1) we deduce that claim (5.3) is valid.

From (5.2), our kernel estimates, and Calderon-Zygmund type arguments we see that $\lim_{\substack{\epsilon\to 0\\ N\to\infty}} [(p_1+p_5)^{2m+1}]$, and the adjoint of this operator are of weak type 1-1. Using Marcinkiewicz interpolation it follows that $\lim_{\substack{\epsilon\to 0\\ N\to\infty}} [(p_1+p_5)^{2m+1}]$ is a bounded operator on $L^r(R^2)$, when $1 \le r < \infty$. It then follows from an argument of Cotlar (see [T, ch 11]) that the maximal operator

$$\sup_{\epsilon,N} |[(p_1+p_5)^{2m+1}]g|(x,t),\ x,t \in R,$$

is bounded on $L^r(R^2)$. Keeping track of the constants we get for c large enough, $i = 0$, and $\eta = \{(1,+)\}$ that

$$||\sup_{\epsilon,N} |[(p_1+p_5)^{2m+1}]g|\ ||_r \le c(r)\beta(1,m,a_1,a_2,c)||g||_r, \tag{5.10}$$

where β is defined as in (0.17).

Next we show that (5.3) holds pointwise almost everywhere in R^2. To do this we follow the same procedure as above. Again the crux of the argument consists in showing that $\lim_{\epsilon\to 0}[p_5]_{\epsilon,N}1$ exists pointwise almost everywhere. To simplify our notation we fix N, recall that $l = 1,\ i = 0, 2, ..., \eta = \{(1,+)\}$, and set

$$S_{\epsilon,\delta}(x,t) = ([p_5]_{\delta,N}1 - [p_5]_{\epsilon,N}1)(x,t), \qquad (x,t) \in R^2,$$

$$S_\delta(x,t) = ([p_5]_{\delta,N}1 - \lim_{\epsilon\to 0}[p_5]_{\epsilon,N}1)(x,t), \quad (x,t) \in R^2,$$

where the limit is in the norm of $L^2(R^2)$ and $0 < \epsilon < \delta$. We show for almost every $(x,t) \in R^2$ that

$$\sup_{0<\epsilon<\delta} |S_\epsilon(x,t)| \le cM(S_\delta)(x,t) + \lambda(x,t,\delta), \tag{5.11}$$

where $|\lambda| \le c(\alpha,i,n)A(1)$, and $\lambda(x,t,\delta) \to 0$ as $\delta \to 0$. Moreover, $M(S_\delta)$ is the one dimensional maximal function of $S_\delta(x,.)$. We observe from (5.7) with $l = 1$, that $S_\delta \to 0$ in $L^2(R^2)$ as $\delta \to 0$. Since

$$(\limsup_{\rho\to 0} S_\rho - \liminf_{\rho\to 0} S_\rho)(x,t) \le 2 \sup_{0<\epsilon<\delta} |S_\epsilon(x,t)|$$

for each $\delta > 0$ it then follows from (5.11), dominated convergence, and the Hardy-Littlewood maximal theorem that $\lim_{\rho\to 0} S_\rho(x,t) = 0$ for almost every $(x,t) \in R^2$, which clearly implies,

$$\lim_{\rho\to 0} [p_5]_{\rho,N}1(x,t) \tag{5.12}$$

exists almost everywhere. To prove (5.11) we observe from (2.24b) and the Tonelli theorem that for almost every $(x_0, t_0) \in R^2$,

$$\int_0^1 \frac{(q(x_0, t_0 + u) - q(x_0, t_0))^2}{u^{1+2\alpha}} du < +\infty.$$

This inequality and (0.4) for q imply that

$$\lim_{s \to t_0} \frac{q(x_0, s) - q(x_0, t_0)}{|s - t_0|^\alpha} = \lim_{s \to t_0} \frac{p_5(x_0, t_0, y, s)}{|s - t_0|^\alpha} = 0. \tag{5.13}$$

Suppose that (x_0, t_0) is a point where (5.13) is true and $|t - t_0| < \epsilon^{1/\alpha}$. Then it is easily shown using (5.13) and (0.4) that

$$|S_{\epsilon,\delta}(x_0, t) - S_{\epsilon,\delta}(x_0, t_0)| \le \lambda_1(x_0, t_0, \delta) \le A(1)c(\alpha, i, n),$$

where $\lambda_1(x_0, t_0, \delta) \to 0$ as $\delta \to 0$. Hence if $I = (t_0 - \epsilon^{1/\alpha}, t_0 + \epsilon^{1/\alpha})$, and $|S_{\epsilon,\delta}|(x_0, t_0) = S_{\epsilon,\delta}(x_0, t_0)$, then

$$\begin{aligned} 2\epsilon^{1/\alpha} S_{\epsilon,\delta}(x_0, t_0) &= \int_I S_\delta(x_0, t)dt - \int_I S_\epsilon(x_0, t)dt + \lambda_1(x_0, t_0, \delta)(2\epsilon^{1/\alpha}) \\ &= J_1 + J_2 + \lambda_1(x_0, t_0, \delta)(2\epsilon^{1/\alpha}). \end{aligned}$$

Clearly,

$$|J_1| \le 2\epsilon^{1/\alpha} M(S_\delta)(x_0, t_0). \tag{5.14}$$

Also, put $\omega(r) = \sup_{|s-t_0| \le r} |p(x_0, s) - p(x_0, t_0)|$, and note for $a > 1$ large that

$$J_2 = \int_I \int_{\{|u| \le a\epsilon^{1/\alpha}\}} \cdots + \int_I \int_{\{|u| > a\epsilon^{1/\alpha}\}} \cdots = J_3 + J_4. \tag{5.15}$$

Now, for $|u| \le a\epsilon^{1/\alpha}$

$$|\int_I p_5(x_0, t, y, t + u)dt| = |\int_I q(x_0, t + u) - q(x_0, t)dt| \le c|u|\omega(2a\epsilon^{1/\alpha}), \tag{5.16}$$

and for $|u| \ge a\epsilon^{1/\alpha}$,

$$|\int_I p_5(x_0, t, y, t + u)dt| \le ca_2\epsilon^{1/\alpha}|u|^\alpha. \tag{5.17}$$

Using (5.16) we obtain from (2.3) with $\lambda = 0,\ \nu = \alpha,\ l = 0$, that

(5.18)

$$|J_3| \le c \int_{-a\epsilon^{1/\alpha}}^{a\epsilon^{1/\alpha}} \int_R \Big(\int_{|x_0-y|}^{\infty} \cdots \int_{\xi_2}^{\infty} |u|^{-[(n+i)\alpha+1]} |\xi_1|^{n-2} \exp\left[\frac{-|\xi_1|^{1/\alpha}}{|u|}\right]$$

$$\cdot\, d\xi_1 \cdots d\xi_i \Big) \Big| \int_I p_5(x_0,t,y,t+u)dt \Big| dudy$$

$$\le c(\alpha,i,n)\omega(2a\epsilon^{\frac{1}{\alpha}}) \Big(\int_0^\infty \int_0^\infty u^\alpha V_{i,0}(z,u) dz du \Big)$$

$$\le c(\alpha,i,n)\omega(2a\epsilon^{\frac{1}{\alpha}}) \Big(\int_0^{a\epsilon^{\frac{1}{\alpha}}} |u|^{-\alpha} du \Big)$$

$$\le c(\alpha,i,n)\omega(2a\epsilon^{\frac{1}{\alpha}})\,(a\epsilon^{\frac{1}{\alpha}})^{1-\alpha}.$$

Using (5.17) we note from (2.14) with M = 1, L = 2, N = a, l = 0, and (2.18 a) that for $i = 2, 4, \ldots,$

(5.19)

$$|J_4|$$

$$\le c a_2 \epsilon^{\frac{1}{\alpha}} \int_{a\epsilon^{\frac{1}{\alpha}}}^{\infty} \int_{x_0-\epsilon}^{x_0+\epsilon} \cdots \int_{\xi_2}^{\infty} |u|^{-[(n+i)\alpha+1]} |\xi_1|^{n-2} \exp\left[\frac{-|\xi_1|^{1/\alpha}}{|u|}\right] d\xi dy du$$

$$\le c a_2 \epsilon^{\frac{1}{\alpha}} \int_a^\infty \mathcal{H}_2(z,u) du \le c(\alpha,i,n) a_2 \epsilon^{\frac{1}{\alpha}} a^{-\alpha/2},$$

where again $d\xi = d\xi_1 d\xi_2 \cdots d\xi_i$ and the last inequality was deduced from (2.15b). If $i = 0$, it is easily checked that (5.18), (5.19), are still true. From (5.13) again, (5.19), (5.18), and (5.15) we find for $a = a(\epsilon)$ properly chosen that

$$|J_2| \le \lambda_2(x_0, t_0, \delta)\epsilon^{1/\alpha}, \tag{5.20}$$

where $\lambda_2 \to 0$ as $\delta \to 0$. From (5.20) and (5.14) we conclude first that (5.11) holds and second that (5.12) is true. Next if $g \in C_0^\infty(R^2)$, we write at (x_0, t_0),

$$[p_5]_{\epsilon,N}(g) = [p_5]_{\epsilon,N}(g - g(x_0,t_0)) + g(x_0,t_0)[p_5]_{\epsilon,N}1.$$

Letting $\epsilon \to 0$ and using the smoothness of g we see that the first term on the right-hand side of this equality has a limit. In view of (5.12) we conclude that $\lim_{\epsilon\to 0} [p_5]_{\epsilon,N}\, g$ exists pointwise for almost every (x,t) when $g \in C_0^\infty(R^2)$. Next from Cotlar's argument and (5.7) with $l = 1$, we note as previously, that $\sup_{\epsilon,N} |[p_5]_{\epsilon,N}\, g|$ is in $L^r(R^2)$ with norm $\le c(r,\alpha,i,n)A(1)\|g\|_r$, $1 < r < \infty$. Using this note and the fact that $C_0^\infty(R^2)$ is dense in $L^r(R^2)$, we conclude that $\lim_{\epsilon\to 0}[p_5]_{\epsilon,N}\, g$ exists pointwise whenever $g \in L^r(R^2)$, $1 < r < \infty$. Finally, existence of $\lim_{\epsilon\to 0} [p^{l-1}p_5]_{\epsilon,N}\, g$ pointwise

, when $l+i$ is odd and $g \in L^r(R^2)$, $1 < r < \infty$, follows from the previous program and induction. We omit the details.

Next, it is easily shown that $\lim_{\epsilon\to 0}[p_1^l]_{\epsilon,N}\, g$ exists pointwise for almost every (x,t) in R^2. Indeed, if $l=0$, then $[p_1^l]_{\epsilon,N}1 \equiv 0$, since i is odd. From this fact, our previous program, and induction we conclude the existence of the above limit. Finally, from existence of these limits, (5.1), and Proposition 2.5 we find

$$\lim_{\epsilon\to 0}\, [(p_1+p_5)^{2m+1}]_{\epsilon,N}\, g(x,t) \tag{5.21}$$

exists for almost every (x,t) when $g \in L^r(R^2)$.

Before we prove Theorem 1 we need to consider some new operators. Let $\Psi_3(x,t,y,s)$ denote the kernel corresponding to $[(p_1+p_5)^{2m+1}]_{\epsilon,N}$ for $i=0$, $l = 2m+1$, $\eta = \{(1,+)\}$, and ϵ, N fixed. Let

$$E = E(\epsilon,N) = \{(y,s) : \epsilon/2 \le |x-y| < \epsilon\} \cup \left\{(y,s) : |s-t| \le \epsilon^{1/\alpha},\ |y-x| \ge \frac{\epsilon}{2}\right\},$$

and put

$$\tilde{\Psi}_3(x,t,y,s) = \Psi_3(x,t,y,s)$$

when $(x,t,y,s) \notin E$ and $\tilde{\Psi}_3 = 0$, otherwise. For $g \in L^r(R^2)$ and $1 < r < \infty$, define

$$T_{\epsilon,N}g(x,t) = \iint_{R^2} \tilde{\Psi}_3(x,t,y,s)g(y,s)dyds.$$

Let $M^1g(x,t)$, $M^2g(x,t)$, denote the Hardy-Littlewood maximal function on R of the functions $g(\cdot,t)$, $g(x,\cdot)$, respectively. From (0.1), (0.4), and a standard argument we find that

$$\begin{aligned} | \quad & T_{\epsilon,N}g \ - [(p_1+p_5)^{2m+1}]g|(x,t) \\ & \le c4^m \iint_E (a_1^{2m+1}|x-y|^{2m+1} + a_2^{2m+1}|s-t|^{\alpha(2m+1)})|s-t|^{-[(2m+n)\alpha+1]} \\ & \cdot \exp\left[\frac{-|x-y|^{1/\alpha}}{|s-t}\right]\ |x-y|^{n-2}g(y,s)dyds \\ & \le \beta(1,m,a_1,a_2,c)M^2(M^1g)(x,t), \end{aligned}$$

where β is defined as in section 1. Using this inequality, the Hardy-Littlewood maximal theorem, and (5.10) we deduce that

$$||\sup_{\epsilon,N} |T_{\epsilon,N}\, g||_r \le c(r)\beta(1,m,a_1,a_2,c)||g||_r, \tag{5.22}$$

$g \in L^r(R^2)$, $1 < r < \infty$. We now let $N \to \infty$. Then

$$\lim_{N\to\infty}\ T_{\epsilon,N}\, g = T_\epsilon\, g,$$

$$\lim_{N\to\infty}\ [(p_1+p_5)^{2m+1}]_{\epsilon,N}\, g = [(p_1+p_5)^{2m+1}]_{\epsilon,\infty} g,$$

pointwise and in the norm of $L^r(R^2)$, as is easily shown, using the fact that q has compact support. We assert that

$$\lim_{\epsilon \to 0} |T_\epsilon g - [(p_1 + p_5)^{2m+1}]_{\epsilon,\infty} g|(x,t) = 0 \tag{5.23}$$

for almost every $(x,t) \in R$ and $g \in C_0^\infty(R^2)$. Using the smoothness of g and (0.1), (0.4), we see that (5.23) is true with g replaced by $g - g(x,t)$. Hence we assume $g = 1$ in (5.23). We note for almost every $(x,t) \in R^2$ that

$$\lim_{y \to x} \frac{p_2(x,t,y,s) - q'(x,t)(y-x)}{y-x} = \lim_{y \to x} \frac{q(y,t) - q(x,t) - q'(x,t)(y-x)}{y-x} = 0 \tag{5.24}$$

since $q(\cdot,t)$ is Lipschitz. Let

$$p_8(x,t,y,s) = p_2 + p_6 - q'(x,t)(y-x),$$

and note that at (x,t,y,s)

$$\begin{aligned} &|(p_2 + p_6\ \ )^{2m+1} - (q'(x,t)(y-x))^{2m+1}| \\ &= |(p_8 + q'(x,t)(y-x))^{2m+1} - (q'(x,t)(y-x))^{2m+1}| \\ &\le c(m)(|p_8|^{2m+1} + |q'(x,t)|^{2m}|p_8|). \end{aligned} \tag{5.25}$$

We note that in the integrals whose integrands are $\Psi_3, \tilde{\Psi}_3$, we can replace $(p_1 + p_5)^{2m+1}$ by $(p_1 + p_5)^{2m+1} - (q'(x,t)(y-x))^{2m+1}$ in order to determine the value of the corresponding operator at 1, since the second term integrates to zero. Doing this, using (5.25), (5.24), (5.13), the fact that $p_1 + p_5 = p_2 + p_6$, and dominated convergence we get (5.23) for $g = 1$. From our earlier remark, we then obtain (5.23). From (5.22), (5.23), and our previous program we conclude that

$$\lim_{\epsilon \to 0} T_\epsilon g(x,t) \tag{5.26}$$

exists for almost every $(x,t) \in R^2$ when $g \in L^r(R)$, $1 < r < \infty$.

From (5.26) and (5.22) we conclude that Theorem 1 is true for Λ_1^m when $n = 2$. To prove Theorem 1 for Λ_1^m and $n = 3, 4, \ldots$, we use the method of rotations. let $\sum$ be the unit sphere in R^{n-1}, $\omega \in \sum$, and f as in section 1. Put

$$\begin{aligned} U_{\epsilon,\omega} g(x,t) = \int_{\{|s-t| \ge \epsilon^{1/\alpha}\}} \int_{\{|\rho| \ge \epsilon\}} [f(x + \rho\omega, s) - f(x,t)]^{(2m+1)} \\ \cdot |s-t|^{-[(2m+n)\alpha+1]} \exp\left[\frac{-\rho^{1/\alpha}}{|s-t|}\right] \rho^{n-2} g(x + \rho\omega, s)\, d\rho\, ds. \end{aligned}$$

for $g \in L^r(R^{n-1} \times R)$, $1 < r < \infty$. Suppose that $x = x^* + \lambda\omega$, where x^* is orthogonal to ω and $\lambda \in R$. We note that if $g_1(v,s) = g(x^* + v\omega, s)$, $v, s \in R$, and $q(v,s) = f(x^* + v\omega, s)$, then

$$U_{\epsilon,\omega} g(x^* + \lambda\omega, t) = T_\epsilon g_1(\lambda, t).$$

Let H denote the vector space orthogonal to ω and let dx^* be $n-2$ dimensional Lebesgue measure on H. From the above equality and (5.22) we see that for $1 < r < \infty$,

(5.27)

$$\int_{\boldsymbol{R}^{n-1}} \int_{\boldsymbol{R}} \sup_{\epsilon} |U_{\epsilon,\omega} g(x,t)|^r dt dx = \int_H \left(\int_{\boldsymbol{R}} \int_{\boldsymbol{R}} \sup_{\epsilon} |T_\epsilon \, g_1|^r (\lambda, t) d\lambda dt \right) dx^*$$

$$\leq c(r) \beta(1, m, a_1, a_2, c)^r \int_H \left(\int_{\boldsymbol{R}} \int_{\boldsymbol{R}} |g|^r (x^* + \lambda\omega, t) d\lambda dt \right) dx^*$$

$$\leq c(r) \beta(1, m, a_1, a_2, c)^r \int_{\boldsymbol{R}^{n-1}} \int_{\boldsymbol{R}} |g|^r (x, t) ds dx.$$

Introducing spherical coordinates, we see also that

$$\Lambda^m_{1,\epsilon} g(x,t) = c \int_{\sum} (U_{\epsilon,\omega} g)(x,t) d\sigma,$$

where σ is surface area on $\sum$. Using this equality, Minkowski's inequality for integrals, and (5.27), we get (0.16) of Theorem 1 for $j = 1$. (0.15) is a consequence of (0.16), (5.26), and dominated convergence. The proof of Theorem 1 for $j = 1$ is now complete. Since the proof for Λ^m_j, $2 \leq j \leq 2n$, is similar, we omit the details.
□

6. References

[A] J. Aguirre, *Multilinear pseudodifferential operators and paraproducts,* Dissertation, Washington University in St. Louis, 1981.

[B1] R.M. Brown, *The method of layer potentials for the heat equation in Lipschitz cylinders,* Amer. J. Math. **111** (1989), 359-379.

[B2] R.M. Brown, *The initial-Neumann problem for the heat equation in Lipschitz cylinders,* Trans. Amer. Math. Soc. **320** (1990), 1-52.

[CM] R.R. Coifman and Y.F. Meyer, *A simplified proof of a theorem by G. David and J.-L. Journé on singular integral operators,* in Probability Theory and Harmonic Analysis, J.A. Chao and W.A. Woyczynski, eds., Marcel Dekker, New York, 1986, 61-65.

[CW1] R.R. Coifman and G. Weiss, *Analyse harmonique non-commutative sur certains espaces homogenes,* Lecture Notes in Math. **242**, Springer-Verlag, Berlin, 1971.

[CW2] R.R. Coifman and G. Weiss, *Extensions of Hardy spaces and their uses in analysis,* Bull. Amer. Math. Soc. **83** (1977), 564-643.

[DJ] G. David and J.-L. Journé, *A boundedness criterion for generalized Calderón-Zygmund operators,* Ann. of Math. **120** (1985), 371-397.

[FR] E.B. Fabes and N.M. Riviére, *Dirichlet and Neumann problems for the heat equation in C^1 cylinders,* Proc. of Symposia in Pure Math. (AMS) **35** (1979), part 2, 179-196.

[Fr1] A. Friedman, *Boundary estimates for second order parabolic equations and their applications,* J. Math. Mech.**7**(1958), 771-791.

[Fr2] A. Friedman, *Partial Differential Equations of Parabolic Type,* Prentice-Hall, Englewood Cliffs, NJ 1964; reprinted by Robert E. Krieger, Malabar, FL 1983.

[JT] B.F. Jones, Jr. and C.C. Tu, *Non-tangential limits for a solution of the heat equation in a two-dimensional Lip_α region,* Duke Math. J. **37** (1970), 243-254.

[KW] R. Kaufman and J.M.G. Wu, *Parabolic measure on domains of class $Lip_{1/2}$,* Compositio Mathematica **65** (1988), 201-207.

[Ke] J.T. Kemper, *Temperatures in several variables: kernel functions, representations, and parabolic boundary values,* Trans. Amer. Math. Soc. **167**(1972), 243-262.

[L] P.G. Lemarie, *Algebres d'opérateurs et semi-groupes de Poisson sur un espace de nature homogène,* Publications Mathématiques d'Orsay, Université de Paris-Sud, Département de Mathématique, Orsay (1984), #84-03.

[LeMu1] J.L. Lewis and M.A.M. Murray, *Regularity properties of commutators and layer potentials associated to the heat equation,* Trans. Amer. Math. Soc.

328(1991), 815-842.

[LeMu2] J.L. Lewis and M.A.M. Murray, *Absolute continuity of parabolic measure,* in Partial Differential Equations with Minimal Smoothness and Applications, IMA Volumes In Mathematics And Its Applications **42**(1992), Springer-Verlag, 173-189.

[LeS] J.L. Lewis and J. Silver, *Parabolic measure and the Dirichlet problem for the heat equation in two dimensions,* Indiana U. Math. J. **37** (1988), 801-839.

[Mu] M.A.M. Murray, *Multilinear singular integrals involving a derivative of fractional order,* Studia Math. **87** (1987), 139-165.

[So] V.A. Solonnikov, *Boundary Value Problems of Mathematical Physics II : On boundary value problems for linear parabolic systems of differential equations of general type,* Proc. of the Steklov Institute of Mathematics **83**(1965; AMS translation 1967).

[St] E.M. Stein, *Singular Integrals and Differentiability Properties of Functions,* Princeton Univ. Press, Princeton, 1970.

[Str] R.S. Strichartz, *Bounded mean oscillation and Sobolev spaces,* Indiana U. Math. J. **29** (1980), 539-558.

[T] A. Torchinsky, *Real variable methods in harmonic analysis,* New York, New York, Academic Press, 1986.

CHAPTER II
The David Buildup Scheme

1. Introduction.

In chapter 1 (see Theorem 1) we showed that certain multilinear singular integral operators were bounded on $L^p(R^{n-1} \times R)$ when $1 < p < \infty$, with estimates on the norm of these operators. The kernel of each operator contained an expression of the form $f(x,t) - f(y,s)$, where f is a function from $R^{n-1} \times R$ into R with

$$(1.1) \qquad |f(x,t) - f(y,t)| \le a_1 |x-y|, \quad x, y \in R^{n-1}, t \in R,$$

$$(1.2) \qquad f(x,t) = I_\alpha(b(x,\cdot))(t), \quad x, y \in R^{n-1}, \ t \in R,$$

where, for each fixed $x \in R^{n-1}$, $b(x,\cdot) \in \mathrm{BMO}(R)$ and

$$(1.3) \qquad ||b(x,\cdot)||_* \le a_2.$$

As in chapter 1 we note that (1.1)-(1.3) imply

$$(1.4) \qquad |f(x,s) - f(x,t)| \le c a_2 |s-t|^\alpha, \quad s, t \in R.$$

For the readers convenience we recall some notation from chapter 1. Let

$$W(Z,\tau) = (4\pi\tau)^{-n/2} \exp\left(-\frac{|Z|^2}{4\tau}\right) \chi_{[0,\infty)}(\tau), (Z,\tau) \in R^{n+1} \setminus \{0\},$$

be the fundamental solution to the heat equation in $R^{n+1} \setminus \{0\}$. Given $\epsilon > 0$, let $W_\epsilon(Z,\tau) = W(Z,\tau)$, when both $|Z| > \epsilon$ and $|\tau| > \epsilon^2$. Otherwise, put $W_\epsilon(Z,\tau) = 0$. If $Z = (z, z_n)$, $z \in R^{n-1}$, $z_n \in R$, let $D \subset R^{n+1}$ be the domain,

$$D = \{(Z,t) : z_n > f(z,t)\}$$

and put $D_t = D \cap (R^n \times \{t\})$ for $t \in R$. The outer unit normal to ∂D_t (considered as a domain in R^n) at a point $(X,t) = (x, f(x,t), t)$ is given by

$$\nu_t(x) = \frac{(\nabla_x f(x,t), -1)}{\sqrt{|\nabla_x f(x,t)|^2 + 1}}, \quad \nabla_x = \left(\frac{\partial}{\partial x_1}, \frac{\partial}{\partial x_2}, \ldots, \frac{\partial}{\partial x_{n-1}}\right).$$

Finally we associate the point $(y,s) \in R^{n-1} \times R$ with the point $(Y,s) = (y, f(y,s), s) \in \partial D$, and identify functions defined on $R^{n-1} \times R$ with functions defined on ∂D.

With this notation in mind, we define the boundary double layer heat potential operator, $\mathcal{L}$, acting on a function $g : \partial D \to R$ as follows:

$$(1.5) \qquad \begin{aligned} \mathcal{L}_\epsilon g(x,t) &= \frac{1}{2} \int_{-\infty}^t \int_{R^{n-1}} \frac{\langle \nu_s(y), (x-y, f(x,t) - f(y,s))\rangle}{(t-s)} \\ &\cdot W_\epsilon(X-Y, t-s) \sqrt{1 + |\nabla_y f(y,s)|^2} g(y,s) dy ds, \end{aligned}$$

where X, $Y \in \partial D$ are as above and

$$\mathcal{L} g(x,t) = \lim_{\epsilon \to 0} \mathcal{L}_\epsilon g(x,t)$$

provided this limit exists. One of our main goals in this chapter is to obtain L^p estimates on the maximal operator (see Corollary 1)

$$\bar{\mathcal{L}}g(x,t) = \sup_{\epsilon} |\mathcal{L}_\epsilon g(x,t)| .$$

To do so we first consider the operators $K_{j,\epsilon}$ defined as follows. For $1 \le j \le 2n$, $\epsilon > 0$, and $(x,t),(y,s) \in R^n$, put

$$H_1 (x,t,y,s) = (4\pi)^{\frac{n}{2}} \frac{f(x,t) - f(y,s)}{|s-t|} W(X-Y, |s-t|)$$

$$= \frac{f(x,t) - f(y,s)}{|s-t|^{\frac{n}{2}+1}} \exp\left[\frac{-|x-y|^2 - |f(x,t) - f(y,s)|^2}{4|s-t|} \right] .$$

For $|x-y| \ge \epsilon$ and $|s-t| \ge \epsilon^2$, let $H_{1,\epsilon}(x,t,y,s) = H_1(x,t,y,s)$, and let

$$H_{j,\epsilon}(x,t,y,s) = (4\pi)^{\frac{n}{2}} \frac{x_{(j-1)} - y_{(j-1)}}{|s-t|} W(X-Y, |s-t|),$$

for $2 \le j \le n$, while

$$H_{j,\epsilon}(x,t,y,s) = \operatorname{sgn}(s-t) H_{j-n,\epsilon}(x,t,y,s), n+1 \le j \le 2n.$$

For all other values of (x,t,y,s), set $H_{j,\epsilon}(x,t,y,s) = 0$. Next for $g \in C_0^\infty(R^{n-1} \times R)$, put

$$K_{j,\epsilon}\, g(x,y) = \int_R \int_{R^{n-1}} H_{j,\epsilon}(x,t,y,s)\, g(y,s)\, dyds,$$

when $1 \le j \le 2n$, and $\epsilon > 0$. The singular integral operators $K_j, 1 \le j \le 2n$, are defined for $g \in C_0^\infty(R^{n-1} \times R)$, by setting

$$K_j\, g(x,t) = \lim_{\epsilon \longrightarrow 0} K_{j,\epsilon}\, g(x,t),$$

when $(x,t) \in R^{n-1} \times R$, and this limit exists. The associated maximal operators, $\bar{K}_j$, $1 \le j \le 2n$, are defined by

$$\bar{K}_j g(x,t) = \sup_{\epsilon > 0} |K_{j,\epsilon} g(x,t)|.$$

In this chapter we prove (in section 5)

Theorem 1. *Let f satisfy (1.1) - (1.3) with $\alpha = \frac{1}{2}$ and have compact support in $R^{n-1} \times R$. For $1 \le j \le 2n$, $1 < p < \infty$, the operator K_j may be extended to a bounded operator on $L^p(R^{n-1} \times R)$ so that for all $h \in L^p(R^{n-1} \times R)$*

$$K_j h(x,t) = \lim_{\epsilon \longrightarrow 0} K_{j,\epsilon} h(x,t) \tag{1.6}$$

exists for almost every $(x,t) \in R^{n-1} \times R$, relative to Lebesgue measure on R^n. Moreover, there exist positive constants $c^j(n,p,a_1,a_2) > 0$, $1 \le j \le 2n$, such that

$$||\bar{K}_j h||_p \le c^j(n,p,a_1,a_2)\, ||h||_p, \tag{1.7}$$

and

$$\lim_{a_1,a_2 \longrightarrow 0} c^1(n,p,a_1,a_2) = \lim_{a_1,a_2, \longrightarrow 0} c^{n+1}(n,p,a_1,a_2) = 0. \tag{1.8}$$

In section 5 we point out that Theorem 1 implies the following corollary :

Corollary 1. *Let $f, \mathcal{L}_\epsilon$, be as in (1.1) - (1.5) . Then for $1 < p < \infty$, $\mathcal{L}_\epsilon$ extends to a bounded operator on $L^p(R^{n-1} \times R)$ and (1.6) - (1.8) hold with K_j replaced by $\mathcal{L}$ and c^j, c^1 by $c(n,p,a_1,a_2)$.*

We note that Corollary 1 will be used in chapter 3 to establish the mutual absolute continuity of parabolic measure and a certain projective Lebesgue measure. To outline the proof of Theorem 1 we first show that this theorem follows from Theorem 1 of chapter 1 with $\alpha = \frac{1}{2}$, when $0 < a_2 < \infty$, and a_1 is sufficiently small, say $0 < a_1 \leq a_0$, where $a_0 = a_0(p,n)$. To do this, observe that

$$\exp\left[\frac{-|x-y|^2 - |f(x,t)-f(y,s)|^2}{4|t-s|}\right]$$

$$(1.9) \qquad = \sum_{m=0}^{\infty} \frac{(-1)^m \,(f(x,t)-f(y,s))^{2m}}{m!\, 4^m |t-s|^m} \cdot \exp[-\frac{|x-y|^2}{4|s-t|}].$$

Using this equality, we write for $h \in L^p(R^{n-1} \times R)$, $1 \leq j \leq 2n$,

$$K_{j,\epsilon}h = \sum_{m=0}^{\infty} \frac{(-1)^m \,\Gamma^m_{j,\epsilon}\, h}{m!\, 4^m},$$

where $\Gamma^m_{j,\epsilon}$ is defined in exactly the same way as $\Lambda^m_{j,\epsilon}$ for $\alpha = 1/2$, except that now $|s-t|$ is replaced by $4|s-t|$ in the exponential term in (0.12) of chapter 1. Changing variables in the integral defining $\Gamma^m_{j,\epsilon}$, we see that Theorem 1 of chapter 1 also holds for this operator with $\alpha = 1/2$. From this theorem we deduce that equality in (1.9) holds in the norm of $L^p(R^{n-1} \times R)$, provided $a_0 = a_0(p,n)$ is small enough. Thus for $1 \leq j \leq 2n$, and $(x,t) \in R^{n-1} \times R$, we have

$$\bar{K}_j h(x,t) \leq \sum_{m=0}^{\infty} \frac{\bar{\Gamma}^m_j h(x,t)}{m!\, 4^m},$$

and so (1.7), (1.8), follow from (0.16) of chapter 1 for a_0 small. Moreover, for each positive integer l,

$$\sup_{\epsilon > 0}\left|\left[K_{j,\epsilon} - \sum_{m=0}^{l} \frac{(-1)^m\, \Gamma^m_{j,\epsilon}}{4^m\, m!}\right] h(x,t)\right| \leq \sum_{m=l+1}^{\infty} \frac{\bar{\Gamma}^m_j h(x,t)}{4^m\, m!},$$

when $(x,t) \in R^{n-1} \times R$ and $1 \leq j \leq 2n$. From this equality and (0.15) of chapter 1 with Γ replacing Λ, we find from the usual $\limsup - \liminf$ argument that for small a_0 and almost every $(x,t) \in R^{n-1} \times R$,

$$(1.10) \qquad \lim_{\epsilon \to 0} K_{j,\epsilon}h(x,t) = K_j h(x,t) = \sum_{m=0}^{\infty} \frac{(-1)^m \Gamma^m_j h(x,t)}{m!\, 4^m},$$

when $1 \leq j \leq 2n$, and $h \in L^p(R^{n-1} \times R)$, $1 < p < \infty$. From (1.10) we conclude that (1.6) of Theorem 1 is true for $a_0 = a_0(p,n) > 0$, small enough.

To prove Theorem 1 for $a_1 > a_0$, $0 < a_2 < \infty$, we shall use a method modeled on the David buildup scheme(see [D1], [J]). The David buildup scheme consists of two parts. First in section 3 (see Lemma 3.1) we show for $d > 0$, $x^* = (x_1^*, ..., x_{n-1}^*) \in R^{n-1}$, $t^* \in R$, that if

$$(1.11) \qquad Q_d(x^*, t^*) = J_1 \times J_2 \times ... \times J_n$$

where $J_i = \{x : |x_i - x_i^*| < (d/2)^\alpha\}$, for $1 \le i \le n-1$, and $J_n = \{t : |t - t^*| < d/2\}$, then $f = f_1$ on a set $E \subset Q_d(x^*, t^*)$ with

$$(1.12) \qquad |Q_d(x^*, t^*)| \le 12|E|,$$

where $f_1 : R^{n-1} \times R \to R$ satisfies (1.1), (1.4), and the operators in Theorem 1 defined relative to f_1 are bounded (see Lemma 4.1) . Here, as in the sequel, $|E|$ denotes the outer Lebesgue n measure of E.

The second part of the David buildup scheme consists in showing that (1.12) and Calderon-Zygmund type estimates on our kernels imply Theorem 1 is true for f. We remark that the construction of f_1 is considerably more complicated than the corresponding construction in [J] , since neither the "rising sun argument" nor a rotation in the spatial coordinates need preserve (1.2)-(1.3). We compensate for this deficiency by proving a key lemma (Lemma 2.1 in section 2) which allows us to replace each of the functions modified in the spatial coordinates, by a function which satisfies (1.2) and (1.3) with a_2 replaced by $\tilde{a}_2 = \tilde{a}_2(a_1, a_2)$. This function also agrees with the modified function on a significant part of $Q_d(x^*, t^*)$.

In the proof of Theorem 1 we were led naturally to consider weaker conditions than (1.2)-(1.3). In order to state these conditions we note as in [St] that (1.2)-(1.3) imply for every $x \in R$ the measure

$$(1.13) \qquad d\mu_x(s, t) = \frac{|f(x, s+t) - f(x, s)|^2}{|t|^{1+2\alpha}} \, dsdt$$

is a Carleson measure on R^2_+, i.e. , there exists c, $0 < c < \infty$, such that

$$(1.14) \qquad \mu_x(\{(s, t) : |s - s_0| < d, \ |t| < d\}) \le c a_2^2 \, d,$$

whenever $s_0 \in R$ and $d > 0$. Also, (1.13)-(1.14) imply (1.2)-(1.3) with a_2 replaced by $c' a_2$, provided $c' > 0$ is large enough . Thus conditions (1.2)-(1.3) are equivalent to (1.13)-(1.14). Now suppose instead of the assumption that (1.13) holds for every $x \in R^{n-1}$, we assume for some $\delta \in (0, 1]$ and every $(x^*, t^*) \in R^{n-1} \times R$, $d > 0$, that

$$(1.15) \qquad \int_{J_n} \frac{(f(x, t) - f(x, s))^2}{|t - s|^{1+2\alpha}} \, ds \le \delta^{-1} a_2^2 \, |J_n|,$$

for (x, t) in a set $E_1 \subset Q_{\frac{d}{2}}(x^*, t^*)$ with

$$(1.16) \qquad |E_1| \ge \delta d^{1+(n-1)\alpha}.$$

Observe from weak type estimates that (1.13)-(1.14) are stronger assumptions than (1.15)-(1.16). In section 5 we show that Lemma 2.1, Theorem 1, and the second part of the David buildup scheme (Lemma 4.2) can be used to prove a stronger theorem:

Theorem 2. *Let f be a function on $R^{n-1} \times R$ which satisfies (1.1), (1.4), and (1.15)-(1.16). Then (0.15), (0.16) of Theorem 1 in chapter 1 remain true if the constants in these inequalities are replaced by a suitably large constant depending on n, p, a_1, a_2, α and δ. If $\alpha = \frac{1}{2}$, then (1.6)-(1.7) of Theorem 1 in this chapter are still true with constants having the same dependence as above.*

Next, in section 5 of this paper we use the David buildup scheme to weaken (1.1). Suppose $f : R^{n-1} \times R \to R$, satisfies instead of (1.1) the condition that $\frac{\partial f}{\partial x_j}$ exists for each j, $1 \le j \le n-1$, in the distributional sense and $\frac{\partial f}{\partial x_j} \in BMO(R^{n-1} \times R)$ on 'rectangles'. That is, if $d > 0$, $(x^*, t^*) \in R^{n-1} \times R$, $Q = \bar{Q}_d(x^*, t^*)$, and

$$m_{j,Q} = m_Q\left(\frac{\partial f}{\partial x_j} \right) = |Q|^{-1} \int_Q \frac{\partial f}{\partial x_j} (x,t) dx dt \, ,$$

then

$$\int_Q \left| \frac{\partial f}{\partial x_j} (x,t) - m_{j,Q} \right| dx dt \le a_1 \, |Q|. \tag{1.17}$$

for $1 \le j \le n$. Let $\zeta(x,t) = (x, f(x,t), t)$, when $(x,t) \in R^{n-1} \times R$, and put $\mathcal{S} = \{\zeta(x,t) : (x,t) \in R^{n-1} \times R\}$. Define σ a positive Borel measure on $R^n \times R$ by

$$\sigma(F) = \int_{\zeta^{-1}(F \cap \mathcal{S})} \sqrt{1 + |\nabla_y f(y,s)|^2} \, dy ds,$$

when F is a Borel subset of $R^n \times R$. Let $X^* = (x_1^*, ..., x_n^*) = (x^*, x_n^*)$, and let $Q_d(X^*, t^*)$ denote the $n+1$ dimensional cube defined as in (1.11) with n replaced by $n+1$ and (x^*, t^*) by (X^*, t^*). We assume for some $\delta > 0$ and every $(X^*, t^*) \in \mathcal{S}$, $d > 0$, that there exists E_1 for which (1.15), (1.16) hold whenever $(x,t) \in E_1$.

Next let W_ϵ be as in (1.5) and put $W_{j,\epsilon}(Z, \tau) = \frac{\partial W_\epsilon}{\partial z_j}(Z, \tau)$, when $1 \le j \le n$, $Z = (z_1, ..., z_n, t)$, and $|Z| \ne \epsilon \ne |\tau|^{\frac{1}{2}}$. If $n+1 \le j \le 2n$, let $W_{j,\epsilon}(Z, \tau) = \operatorname{sgn} \tau \, W_{j-n,\epsilon}(Z, \tau)$. Let $L^p(\sigma)$, $1 \le p \le \infty$, denote the space of σ measurable functions g with

$$||g||_{p,\sigma} = \left[\int |g|^p \, d\sigma \right]^{\frac{1}{p}} < \infty.$$

If $g \in C_0^\infty(R^n \times R)$, and $1 \le j \le 2n$, set

$$P_{j,\epsilon} \, g(X,t) = \int W_{j,\epsilon}(X - Y, t - s) \, g(Y,s) \, d\sigma(Y,s)$$

and

$$\bar{P}_j \, g(X,t) = \sup_{\epsilon > 0} | P_{j,\epsilon} \, g(X,t) |,$$

when $(X,t) \in R^n \times R$. In section 5 we prove

Theorem 3. *Let f be a function on $R^{n-1} \times R$ which satisfies (1.4), and (1.15)- (1.17). Then (1.8), (1.9), of Theorem 1 are valid with $L^p(R^{n-1} \times R)$ replaced by $L^p(\sigma)$ and K_j by P_j. The norms of these operators can be estimated by constants having the same dependence as in Theorem 2. Also the conclusion of Corollary 1 remains valid with these replacements.*

In section 6 we state some further results (Theorem 4) concerning parabolic analogues of ω regular surfaces in the sense of David [D1].

2. Key Lemma.

This section is devoted to a Lemma which is an essential ingredient in the proofs of Theorems 1-3. We begin with some notation and terminology. In the sequel we identify $R^{n-1} \times R$ and R^n. Also c will denote a positive constant which may only depend on α and n, not necessarily the same at each occurrence. Let $\bar{F}$ denote the closure of the set F. If $I \subset R$ let $l(I)$ denote the one dimensional outer Lebesgue measure of I. If I is an interval let βI denote the interval with the same center as I and $l(\beta I) = \beta l(I)$. Fix α, $0 < \alpha < 1$, and let $Q_d(x^*, t^*)$ be as in (1.11). If S is a subset of R^n and $\phi : S \to R$ satisfies

$$|\phi(x,t) - \phi(y,t)| \leq c|x-y|, \text{ for all } (x,t),(y,t) \in S$$

(compare (1.1)) then we say that ϕ is " uniformly Lipschitz in the sense of (1.1)." Finally we introduce a nonisotropic distance function : for $(x,t),(y,s) \in R^n$, define

$$d((x,t),(y,s)) = |x-y| + |s-t|^\alpha.$$

Then, if F_1, F_2 are two bounded subsets of R^n, we set

$$d(F_1, F_2) = \inf\{d((x,t),(y,s)) : (x,t) \in F_1, (y,s) \in F_2\},$$

$$\operatorname{diam}(F_1) = \sup\{d((x,t),(y,s)) : (x,t),(y,s) \in F_1\}.$$

Our key lemma is as follows.

Lemma 2.1 *Let $\phi_1, \phi_2 : R^n \to R$ be two functions supported in $\bar{Q}_d(x^*,t^*)$ which are uniformly Lipschitz in the sense of (1.1). Let $b_1, b_2, ..., b_n, e_1, ..., e_{n-1}$ be constants, $b = \sum_{j=1}^{n-1} |b_j|$, $e = \sum_{j=1}^{n-1} |e_j|$. satisfying*

$$b_j \leq \frac{\partial \phi_1}{\partial x_j}(x,t) \leq e_j, \; 1 \leq j \leq n-1, \tag{2.1}$$

$$-b-e \leq \frac{\partial \phi_2}{\partial x_j}(x,t) \leq b+e, \; 1 \leq j \leq n-1, \tag{2.2}$$

for almost every $(x,t) \in Q_d(x^*,t^*)$, and

$$|\phi_i(x,t) - \phi_i(x,s)| \leq b_n\,|s-t|^\alpha, \; i = 1,2, \tag{2.3}$$

for $(x,t),(x,s) \in \bar{Q}_d(x^*,t^*)$. *Suppose further that* F *is a nonempty closed subset of* $\bar{Q}_{\frac{d}{2}}(x^*,t^*)$, *with*

$$\int_{J_n} \frac{(\phi_2(x,t)-\phi_2(x,s))^2}{|s-t|^{1+2\alpha}}\, ds \le b_n^2, \tag{2.4}$$

for $(x,t) \in F$, *while*

$$\phi_2 = \phi_1 \text{ on } F. \tag{2.5}$$

Then there is a positive constant $c_2 = c_2(\alpha, n)$, *such that for every* $\gamma \in (0,1)$, *there exists a compactly supported function* $\phi : R^n \to R$ *which is uniformly Lipschitz in the sense of (1.1), and*

$$b_j - c_2\gamma(b+e) < \frac{\partial \phi}{\partial x_j}(x,t) < e_j + c_2\gamma(b+e), 1 \le j \le n-1, \tag{2.6}$$

for almost every $(x,t) \in R^n$. *Moreover,*

$$|\phi(x,s)-\phi(x,t)| \le c\, b_n' \, |s-t|^{\alpha},\ (x,t),(x,s) \in R^n, \tag{2.7}$$

and, whenever $J \subset R$ *is an interval,*

$$\int_J \int_J \frac{(\phi(x,t)-\phi(x,s))^2}{|s-t|^{1+2\alpha}}\, ds dt \le c(b_n')^2 l(J), \tag{2.8}$$

for $(x,t) \in R^n$, *where*

$$b_n' = \max\{b_n,\ b+e,\ b_n^{\frac{1}{\alpha}}[\gamma(b+e)]^{1-\frac{1}{\alpha}}\}.$$

Finally,

$$\phi = \phi_1 = \phi_2 \quad \text{on } F. \tag{2.9}$$

Intuitively Lemma 2.1 states that we can weld the " best parts " of ϕ_1, ϕ_2 on F to form a globally-defined function with desirable properties. In particular note that (2.8) implies the measure μ_x in (1.13), defined relative to ϕ (instead of f), is a Carleson measure on R^n_+, with constants independent of $x \in R$. Thus (1.2), (1.3), hold for ϕ with a_2 replaced by $c\, b_n'$.

Proof. We begin the proof of Lemma 2.1 with a nonisotropic Whitney decomposition of $\bar{Q}_d(x^*,t^*) \setminus F$. We claim that there exists a family $\{Q_i\}_1^\infty$ of nonisotropic rectangles, with sides parallel to the axes and disjoint interiors, having the following properties : Q_i has its center at a point (y_i, t_i) and sidelength r_i in the x_j direction ($1 \le j \le n-1$) and r_i' in the t direction . Moreover,

$$(r_i')^{\alpha} \le r_i \le c(r_i')^{\alpha},\ i = 1, \ldots \tag{2.10}$$

$$c d(Q_i, F) < \text{diam}\,(Q_i) < \frac{1}{100\sqrt{n-1}} d(Q_i, F),\ i = 1, 2, \ldots, \tag{2.11}$$

and $\bar{Q}_d(x^*,t^*) \setminus F = \cup Q_i$. The construction of $\{Q_i\}_1^\infty$ follows from a modification of the argument in [S, ch. 6], in which the bisection of the axes proceeds more slowly in the space variables than in the time variable. Specifically, $\bar{Q}_d(x^*, t^*)$ is decomposed into a succession of meshes $G(k), k = 1, 2, \ldots$, where each 'rectangle' in the mesh $G(k)$ has sidelength $2^{-k}d$ in the t direction and $2^{-[k\alpha]}2^{1-\alpha}d^{\alpha}$ in the x direction (here, [.] denotes the greatest integer function); the estimate (2.10) follows easily from this. The rectangles $\{Q_i\}$ are then selected from the meshes $G(k)$ in a manner analogous to [S, ch 6] , so that (2.11) is satisfied. It is clear from the construction, and from (2.10), that Q_i is almost, but not quite, the same as the nonisotropic cube $Q_{r_i'}(y_i, t_i)$.

Next let $0 \le \psi_1 \in C_0^\infty([-2, 2])$ with $\psi_1 > 0$ on $[-1, 1]$ and $\int_R \psi_1 dx = 1$. Put $\psi_2(x) = \psi_1(|x|)$, when $x \in R^{n-1}$. Let

$$v(x,t) = \sum_k \psi_1(\tfrac{t-t_k}{r_k'})\, \psi_2(\tfrac{x-y_k}{r_k\sqrt{n-1}}),$$

$$v_i(x,t) = v(x,t)^{-1}\, \psi_1(\tfrac{t-t_i}{r_i'})\, \psi_2(\tfrac{x-y_i}{r_i\sqrt{n-1}}) .$$

Then clearly, $\{v_i\}$ is a C^∞ partition of unity for $\bar{Q}_d(x^*, t^*) \setminus F$. Moreover, from (2.10), (2.11) we see for fixed k that if

$$S = \{i : \ \text{supp } v_i \cap \bar{Q}_k\} \ne \emptyset,$$

then

$$\text{card } S \le c, \tag{2.12}$$

$$c^{-1} r_l \le r_k \le c r_l, \ \ l \in S. \tag{2.13}$$

We extend the definition of ϕ_1 to $\bar{J}_1 \times \ldots \times \bar{J}_{n-1} \times 2\bar{J}_n$, by requiring for fixed $x \in \bar{J}_1 \times \ldots \times \bar{J}_{n-1}$ that $\phi_1(x, .)$ is constant on each component of $\bar{J}_1 \times \ldots \times \bar{J}_{n-1} \times (2\bar{J}_n \setminus J_n)$.

Let $\epsilon = \left[\dfrac{(b + c)\gamma}{b_n} \right]^{\frac{1}{\alpha}}$, and

$$g_k(x,t) = (\epsilon r_k')^{-1} \int_R \phi_1(x,\tau)\psi_1(\frac{t-\tau}{\epsilon r_k'})\, d\tau$$

whenever $(x,t) \in \bar{Q}_d(x^*, t^*)$, and define $\tilde{\phi}$ on $\bar{Q}_d(x^*, t^*)$ by

$$\tilde{\phi}(x,t) = \begin{cases} \sum_k (g_k v_k)(x,t), & (x,t) \in \bar{Q}_d(x^*, t^*) \setminus F, \\ \phi_1(x,t), & (x,t) \in F. \end{cases}$$

We observe from (2.1) for $1 \le j \le n-1$ and $(x,t) \in \bar{Q}_d(x^*, t^*) \setminus F$ that

$$\begin{aligned} \frac{\partial \tilde{\phi}}{\partial x_j} &= \sum_k \frac{\partial g_k}{\partial x_j} v_k + \sum_k g_k \frac{\partial v_k}{\partial x_j} \\ &\le e_j \sum_k v_k + \sum_k (g_k - \phi_1) \frac{\partial v_k}{\partial x_j} \qquad (2.14) \\ &= e_j + \sum_k (g_k - \phi_1) \frac{\partial v_k}{\partial x_j} . \end{aligned}$$

Moreover, from (2.3), (2.10), and the definition of ϵ, we deduce for $(x,t) \in \bar{Q}_d(x^*,t^*)$, that

(2.15)
$$
\begin{aligned}
|g_k(x,t) - \phi_1(x,t)| &\leq (\epsilon r_k')^{-1} \int_{\mathbf{R}} |\phi_1(x,\tau) - \phi_1(x,t)| \psi_1\left(\frac{t-\tau}{\epsilon r_k'}\right) d\tau \\
&\leq c b_n (\epsilon r_k')^{\alpha} \leq c\gamma(b+e) r_k .
\end{aligned}
$$

Using (2.15) in (2.14), together with (2.12), (2.13), and the fact that $\left|\frac{\partial v_k}{\partial x_j}\right| \leq c r_k^{-1}$, we get for $1 \leq j \leq n-1$ that

$$
\frac{\partial \tilde{\phi}}{\partial x_j} \leq e_j + c\gamma (b+e), \tag{2.16}
$$

whenever $(x,t) \in Q_d(x^*,t^*) \setminus F$, which is the right hand inequality in (2.6) with ϕ replaced by $\tilde{\phi}$. The left hand inequality in (2.6) for $\tilde{\phi}$ is proved similarly when $(x,t) \in Q_d(x^*,t^*) \setminus F$. Next, from (2.12), (2.13), and (2.15) , we note that for $(x,t) \in \bar{Q}_k$,

(2.17)
$$
\begin{aligned}
|\tilde{\phi}(x,t) - \phi_1(x,t)| &= \left|\sum_l (g_l - \phi_1)(x,t) v_l(x,t)\right| \\
&\leq c \sum_{l \in S} |g_l - \phi_1|(x,t) \leq c\gamma(b+e) r_k .
\end{aligned}
$$

Using (2.17), the fact that ϕ_1 is uniformly Lipschitz in the sense of (1.1), and an argument similar to the one used in proving (2.16), we conclude that $\tilde{\phi}$ is uniformly Lipschitz in the sense of (1.1). From (2.1) and (2.6) on $\bar{Q}_d(x^*,t^*) \setminus F$, it follows that (2.6) is valid in $\bar{Q}_d(x^*,t^*)$ for $\tilde{\phi}$. Note by the definition of $\tilde{\phi}$ that (2.9) is true with ϕ replaced by $\tilde{\phi}$.

To prove (2.7) for $\tilde{\phi}$, observe from (2.3), (2.12), and (2.13) that for $(x,t) \in \bar{Q}_k$ and $l \in S$,

$$
\begin{aligned}
\left|\frac{\partial g_l}{\partial t}(x,t)\right| &= \left|(\epsilon r_l')^{-1} \int_{\mathbf{R}} [\phi_1(x,\tau) - \phi_1(x,t)] \frac{d}{dt} \psi_1\left(\frac{t-\tau}{\epsilon r_l'}\right) d\tau\right| \\
&\leq c b_n (\epsilon r_k')^{\alpha-1} \leq c b_n^{\frac{1}{\alpha}} [\gamma(b+e)]^{(1-\frac{1}{\alpha})} (r_k')^{\alpha-1}.
\end{aligned}
$$

This inequality, (2.10), (2.12), (2.13), (2.15), and the fact that $\left|\frac{\partial v_k}{\partial t}\right| \leq c(r_k')^{-1}$ imply at $(x,t) \in \bar{Q}_k$ that

(2.18)
$$
\begin{aligned}
\left|\frac{\partial \tilde{\phi}}{\partial t}\right| &= \left|\sum_i \frac{\partial g_i}{\partial t} v_i + \sum_i (g_i - \phi_1) \frac{\partial v_i}{\partial t}\right| \\
&\leq c(r_k')^{\alpha-1} \{ b_n^{\frac{1}{\alpha}} [\gamma(b+e)]^{1-\frac{1}{\alpha}} + \gamma(b+e) \} \\
&\leq c\, b_n' (r_k')^{\alpha-1}.
\end{aligned}
$$

Fix $(x,t) \in \bar{Q}_k$. If $(x,s) \in \bar{Q}_l$ and $\bar{Q}_l \cap \bar{Q}_k \neq \emptyset$, then from (2.13) and (2.18) it is easily seen that

$$|\tilde{\phi}(x,t) - \tilde{\phi}(x,s)| \leq c\, b_n' |s-t|^\alpha. \tag{2.19}$$

Otherwise, either $(x,s) \in \bar{Q}_j$, and

$$c\,|s-t| \geq \max(r_k', r_j')$$

or $(x,s) \in F$ and $c\,|s-t| \geq r_k'$. In either case it follows from (2.17), (2.10), (2.11), and (2.3) that (2.19) holds when $(x,t) \in Q_k$ and $(y,s) \in \bar{Q}_d(x^*,t^*)$. Letting Q_k vary and using (2.3) , (2.9) once again, we obtain (2.19) in $\bar{Q}_d(x^*,t^*)$.

Next we prove (2.8) for $\tilde{\phi}$ when $(x,t) \in \bar{Q}_d(x^*,t^*)$ and $J \subset \bar{J}_n$. The argument we give is a two dimensional version of the argument in [LM]. For fixed $x \in \bar{J}_1 \times \bar{J}_2 \times \ldots \bar{J}_{n-1}$, let

$$H = H(x) = \{t \in \bar{J}_n : (x,t) \in F\},$$

and suppose that $J_n \setminus H = \cup I_j$, where $\{I_j\}$ are pairwise disjoint open intervals with the property that for each j there exists $l = l(j)$ with

$$\bar{I}_j = \{(x,t) : t \in Q_l\}.$$

From (2.18), (2.10), and (2.11), we deduce for a given closed interval $J \subset \bar{J}_n$ that

$$\int_{4I_k \cap J} \int_{4I_k \cap J} \frac{(\tilde{\phi}(x,s) - \tilde{\phi}(x,t))^2}{|s-t|^{1+2\alpha}}\, ds dt \leq c\,(b_n')^2\, l(4I_k \cap J), \tag{2.20}$$

whenever $I_k \in \{I_j\}$. Also from (2.7) for $\tilde{\phi}$ if $l(J) \geq l(I_k)$,

(2.21)

$$\int_{I_k \cap J} \int_{J \setminus 4I_k} \frac{(\tilde{\phi}(x,s) - \tilde{\phi}(x,t))^2}{|s-t|^{1+2\alpha}}\, ds dt \leq c(b_n')^2 \int_{I_k \cap J} \left(\int_{J \setminus 4I_k} |s-t|^{-1} ds \right) dt$$

$$\leq c\,(b_n')^2\, l(I_k \cap J) \log\left(\frac{2l(J)}{l(I_k)} \right) \leq c\,(b_n')^2\, l(J).$$

If $l(J) < l(I)$, the above inequality is trivially true since in this case the left hand side of (2.21) is zero.

Now

$$
\begin{aligned}
&\int_J \int_J \frac{(\tilde{\phi}(x,s)-\tilde{\phi}(x,t))^2}{|s-t|^{1+2\alpha}}\, dsdt \\
&= \int_{H\cap J} \int_{H\cap J} \frac{(\tilde{\phi}(x,s)-\tilde{\phi}(x,t))^2}{|s-t|^{1+2\alpha}}\, dsdt \\
&+2\sum_k \int_{I_k\cap J} \int_{H\cap J} \frac{(\tilde{\phi}(x,s)-\tilde{\phi}(x,t))^2}{|s-t|^{1+2\alpha}}\, dsdt \qquad (2.22)\\
&+\sum_{k,j} \int_{I_k\cap J} \int_{I_j\cap J} \frac{(\tilde{\phi}(x,s)-\tilde{\phi}(x,t))^2}{|s-t|^{1+2\alpha}}\, dsdt \\
&= T_1 + T_2 + T_3.
\end{aligned}
$$

Since $\tilde{\phi} = \phi_2$ on H, we find from (2.4) that

$$
T_1 \le c\, b_n^2\, l(J). \qquad (2.23)
$$

We observe from properties (2.10), (2.11), of $\{Q_j\}$ that there exists $(z_k, \tau_k) \in F$ with

$$
c^{-1}\, l(I_k)^\alpha \le |z_k - x| + |t-\tau_k|^\alpha \le c\, l(I_k)^\alpha, \qquad (2.24)
$$

whenever $t \in I_k \in \{I_j\}$. Then from (2.2) , (2.3), (2.6), (2.7), (2.9), and (2.24), we deduce

$$
\begin{aligned}
&(\phi_2(x,t)-\tilde{\phi}(x,t))^2 \\
&\le 2(\phi_2(x,t)-\phi_2(z_k,\tau_k))^2 + 2(\tilde{\phi}(x,t)-\tilde{\phi}(z_k,\tau_k))^2 \qquad (2.25)\\
&\le c(b_n')^2\, l(I_k)^{2\alpha},
\end{aligned}
$$

whenever $t \in I_k \in \{I_j\}$.

Next put $M = \{I_k : I_k \subset J\}$. Since $\{I_j\}$ have disjoint interiors, we see that $\{I_k : I_k \cap J \ne \emptyset\} \setminus M$ consists of at most two intervals. From this fact, (2.20),

(2.21), (2.25), (2.9), and (2.4), we find that

$$
\begin{aligned}
T_2 &\leq c(b_n')^2 l(J) + c \sum_{I_k \in M} \int_{I_k} \left[\int_{(H\setminus 4I_k)\cap J} \frac{(\tilde{\phi}(x,s) - \tilde{\phi}(x,t))^2}{|s-t|^{1+2\alpha}} \, ds \right] dt \\
&\leq c(b_n')^2 \, l(J) + c \int_J \left[\int_{J\cap H} \frac{(\phi_2(x,s) - \phi_2(x,t))^2}{|s-t|^{1+2\alpha}} \, ds \right] dt \\
&\quad + c \sum_{I_k \in M} \int_{I_k} (\phi_2(x,t) - \tilde{\phi}(x,t))^2 \left[\int_{\{|s-t|>2l(I_k)\}} \frac{ds}{|s-t|^{1+2\alpha}} \right] \\
&\leq c(b_n')^2 l(J) + c b_n^2 l(J) + (b_n')^2 \Big(\sum_{I_k \in M} l(I_k) \Big) \\
&\leq c(b_n')^2 l(J).
\end{aligned} \tag{2.26}
$$

Let $M(k) = \{I_j \in M : l(I_j) \leq l(I_k)\}$. Then from (2.20), (2.21), we have

$$
\begin{aligned}
T_3 &\leq c(b_n')^2 \, l(J) + c \sum_{I_k \in M} \sum_{I_j \in M(k)} \int_{I_k} \int_{I_j \setminus 4I_k} \frac{(\tilde{\phi}(x,s) - \tilde{\phi}(x,t))^2}{|s-t|^{1+2\alpha}} \, dsdt \\
&\leq c(b_n')^2 \, l(J) + c \sum_{I_k \in M} \sum_{I_j \in M(k)} \int_{I_k} \int_{I_j \setminus 4I_k} \frac{(\phi_2(x,s) - \phi_2(x,t))^2}{|s-t|^{1+2\alpha}} \, dsdt \\
&\quad + c \sum_{I_k \in M} \sum_{I_j \in M(k)} \int_{I_k} \int_{I_j \setminus 4I_k} \frac{(\tilde{\phi}(x,t) - \phi_2(x,t))^2}{|s-t|^{1+2\alpha}} \, dsdt \\
&\quad + c \sum_{I_k \in M} \sum_{I_j \in M(k)} \int_{I_k} \int_{I_j \setminus 4I_k} \frac{(\tilde{\phi}(x,s) - \phi_2(x,s))^2}{|s-t|^{1+2\alpha}} \, dsdt \\
&= c(b_n')^2 \, l(J) + N_1 + N_2 + N_3 \, .
\end{aligned} \tag{2.27}
$$

We also deduce

$$
\begin{aligned}
N_1 &\leq c\sum_{k,j} \int_{I_k} \int_{I_j\setminus 4I_k} \frac{(\phi_2(x,s)-\phi_2(z_k,s))^2}{|s-t|^{1+2\alpha}}\, dsdt \\
&+ c\sum_{k,j} \int_{I_k} \int_{I_j\setminus 4I_k} \frac{(\phi_2(z_k,s)-\phi_2(z_k,\tau_k))^2}{|s-t|^{1+2\alpha}}\, dsdt \\
&+ c\sum_{k,j} \int_{I_k} \int_{I_j\setminus 4I_k} \frac{(\phi_2(z_k,\tau_k)-\phi_2(x,t))^2}{|s-t|^{1+2\alpha}}\, dsdt \\
&= W_1 + W_2 + W_3,
\end{aligned}
\tag{2.28}
$$

where the sum is taken over all k, j with I_k, $I_j \in M$ and $\{(z_k,\tau_k)\}$ as in (2.24). From (2.2),(2.3), and (2.24), we obtain

$$
\begin{aligned}
W_1 + W_3 &\leq c\,(b_n')^2 \sum_k \int_{I_k} \left(\int_{R\setminus 4I_k} \frac{l(I_k)^{2\alpha}}{|s-t|^{1+2\alpha}}\, ds \right) dt \\
&\leq c\,(b_n')^2\, l(J).
\end{aligned}
$$

Moreover, since $(z_k,\tau_k) \in F$ and $c\,|s-t| \geq |s-\tau_k|$ when $s \in R\setminus 4I_k$, $t \in I_k$, we get from (2.4) that $W_2 \leq c\,(b_n')^2\, l(J)$. Using these estimates in (2.28), we conclude

$$
N_1 \leq c\,(b_n')^2\, l(J). \tag{2.29}
$$

To estimate N_2, we use (2.25) to obtain

$$
\begin{aligned}
N_2 &\leq c\,(b_n')^2 \sum_{I_k\in M} \int_{I_k} \left(\int_{R\setminus 4I_k} \frac{l(I_k)^{2\alpha}}{|s-t|^{1+2\alpha}}\, ds \right) dt \\
&\leq c\,(b_n')^2\, l(J).
\end{aligned}
\tag{2.30}
$$

To estimate N_3 we note that the inside integral in this expression is zero unless I_k lies at least $\frac{1}{2}l(I_j)$ from I_j, since $l(I_j) \leq l(I_k)$. Using this fact, (2.25), and interchanging the order of summation we get

$$
\begin{aligned}
N_3 &\leq c\,(b_n')^2 \sum_{I_j\in M} \int_{I_j} \left(\int_{R\setminus \frac{3}{2}I_j} \frac{l(I_j)^{2\alpha}}{|s-t|^{1+2\alpha}}\, ds \right) dt \\
&\leq c\,(b_n')^2\, l(J).
\end{aligned}
\tag{2.31}
$$

From (2.27) and (2.29)-(2.31), we have

$$
T_3 \leq c\,(b_n')^2\, l(J). \tag{2.32}
$$

In view of (2.32), (2.26), (2.23), and (2.22), we conclude that (2.8) is true when J is a closed interval and $J \subset J_n$.

Next we extend $\tilde{\phi}$ to a function on R^n (also denoted $\tilde{\phi}$) as follows: first define $\tilde{\phi}$ on $R \times \bar{J}_2 \times ... \times \bar{J}_{n-1} \times 2\bar{J}_n$ by requiring that $\phi(\cdot, x_2, ..., x_{n-1}, t)$ be constant on each component of $R \setminus J_1$, whenever

$$(x_2, ..., x_{n-1}, t) \in \bar{J}_2 \times ... \times \bar{J}_{n-1} \times 2\bar{J}_n .$$

Second, if $n \geq 3$ extend $\tilde{\phi}$ to $R \times R \times ... \times \bar{J}_{n-1} \times 2\bar{J}_n$ by requiring that $\phi(x_1, \cdot, x_3, ..., x_n, t)$ be constant on each component of $R \setminus J_2$ whenever

$$(x_1, x_3, ..., x_n, t) \in R \times \bar{J}_3 \times ... \times \bar{J}_{n-1} \times 2\bar{J}_n .$$

If $n = 2$ extend $\tilde{\phi}$ from $R \times 2\bar{J}_2$ by requiring that $\tilde{\phi}(x, \cdot)$ be constant on each component of $R \setminus 2J_2$, whenever $x \in R$. Continuing in this manner we get $\tilde{\phi}$ defined on R^n. Since $\tilde{\phi}$ satisfies (2.6), (2.7) at points in $\bar{Q}_d(x^*, t^*)$, it is easily checked that (2.6), (2.7) hold at points of R^n. Also, arguing as in [LM2, Lemma 3] , we deduce that (2.8) holds for c sufficiently large.

Finally, let $\beta \in C_0^\infty[-2, 2]$ with $0 \leq \beta \leq 1$ and $\beta \equiv 1$ on $[-1, 1]$. Given $N > 0$ and large put

$$\phi(x, t) = \beta(\frac{|x - x^*|}{N})\beta(\frac{t - t^*}{N})\tilde{\phi}(x, t) ,$$

when $(x, t) \in R^{n-1} \times R$. Clearly ϕ has compact support. We conclude from our earlier work that for N large (depending upon $||\tilde{\phi}||_\infty$, $b + e, \gamma, b_n$) Lemma 2.1 is true. □

3. The David Buildup Scheme: Part 1 .

Let f satisfy (1.1) - (1.3) and have compact support in R^n . Given $d > 0$ let $\bar{Q}_d(x^*, t^*)$ be as in (1.11) . In this section we use the David buildup scheme to prove

Lemma 3.1 *For* f, $\bar{Q}_d(x^*, t^*)$ *as above suppose* $\mu > 0$ and $l \in \{1, 2, ..., n-1\}$ satisfy

$$\mu = || \frac{\partial f}{\partial x_l} ||_\infty = \max_{1 \leq j \leq n-1} || \frac{\partial f}{\partial x_j} ||_\infty .$$

Then there exists a closed set $E \subset \bar{Q}_d(x^*, t^*)$ *and functions* $f_1, f_2 : R^n \to R$ *satisfying (1.1), with* a_1 *replaced by* ca_1, *such that*

$$f_1 = f \text{ on } E, \tag{3.1}$$

$$|E| \geq \frac{1}{12} \mid \bar{Q}_d(x^*, t^*) \mid . \tag{3.2}$$

Moreover, f_1 *is defined by*

$$f_1(x, t) = -\sin\theta \, \bar{x}_l + \cos\theta \, f_2(\bar{x}, t) \tag{3.3}$$

where $(\bar{x}, t) \in R^n$,

$$\theta = \pm \frac{1}{2} [\tan^{-1}(\frac{101\mu}{100}) - \tan^{-1}(\frac{91\mu}{100})]$$

and x differs from $\bar{x}$ in a single coordinate, expressed as a function of $(\bar{x}, t)$ by

$$x_l(\bar{x}, t) = \cos\theta\,\bar{x}_l + \sin\theta\, f_2(\bar{x}, t), \tag{3.4}$$

$$x_j = \bar{x}_j\,, 1 \le j \le n-1,\, j \ne l\,.$$

The function f_2 has compact support in R^n and

$$\| \frac{\partial f_2}{\partial x_l} \|_\infty \le (\frac{24}{25} + \frac{1}{50n})\mu\,, \tag{3.5 a}$$

$$\| \frac{\partial f_2}{\partial x_j} \|_\infty \le \| \frac{\partial f}{\partial x_j} \|_\infty + \frac{\mu}{50n}\,,\ 1 \le j \le n-1,\ j \ne l. \tag{3.5b}$$

Finally , f_1 satisfies (1.4), and f_2 satisfies (1.2) - (1.4) with a_2, b replaced by $\tilde{a}_2, \tilde{b}$, where

$$\tilde{a}_2 = c\,\hat{a}_2\,[\log(2 + \hat{a}_2)]^{\frac{1}{2\alpha}},$$

$$\hat{a}_2 = \max\{\, a_1,\, a_2,\, a_2^{\frac{1}{\alpha^2}}\, \mu^{1-\frac{1}{\alpha^2}} \}$$

Proof. We note from Lemma 3.1 that $f = f_1$ on a significant portion (E) of $\bar{Q}_d(x^*, t^*)$. Also, f_1 is related to a "good" function f_2 by (3.3) with

$$\begin{aligned} |||\nabla_x f_2||| &= \sum_1^{n-1} \| \frac{\partial f_2}{\partial x_j} \|_\infty \\ &\le (\frac{24}{25} + \frac{1}{50n})\mu + \sum_{j \ne l} \| \frac{\partial f}{\partial x_j} \|_\infty + \frac{(n-2)\mu}{50n} \\ &\le \frac{49}{50}\,\mu + \sum_{j \ne l} \| \frac{\partial f}{\partial x_j} \|_\infty \\ &\le [1 - \frac{1}{50(n-1)}]|||\nabla_x f|||, \end{aligned} \tag{3.6}$$

thanks to (3.5). Lemma 3.1 , (3.6), and the second part of the David buildup scheme will enable us to prove Theorem 1 by successively reducing back to the case of small Lipschitz norm.

To begin the construction of f_1, f_2, we assume, as we may, that $l = 1$. Given $(x_2, \ldots, x_{n-1}, t) \in \bar{J}_2 \times \ldots \times \bar{J}_n$, let

$$\hat{f}(z, x_2, \ldots, x_{n-1}, t) = \max_{z \le y \in J_1} [\, f(y, x_2, \ldots, t) - \frac{9}{10}\,\mu y] + \frac{9}{10}\,\mu z,$$

$$\bar{f}(z, x_2, \ldots, x_{n-1}, t) = \min_{z \le y \in J_1} [f(y, x_2, \ldots, t) + \frac{9}{10}\,\mu y] - \frac{9}{10}\,\mu z\,,$$

when $z \in \bar{J}_1$. We note from the rising sun lemma [Z, p 31] that

$$O_1 = O_1(x_2, ..., x_{n-1}, t) = \{z : \hat{f}(z, x_2, ..., x_{n-1}, t) > f(z, x_2, ...x_{n-1}, t)\} \cap J_1,$$

$$O_2 = O_2(x_2, ..., x_{n-1}, t) = \{z : \bar{f}(z, x_2, ..., x_{n-1}, t) < f(z, x_2, ..., x_{n-1}, t)\} \cap J_1,$$

are open in R . Furthermore, $\hat{f}(\cdot, x_2, ..., x_{n-1}, t)$ and $\bar{f}(\cdot, x_2, ..., x_{n-1}, t)$ are continuous on $\bar{J}_1$ and linear on each component of O_1, O_2, respectively. Therefore if (a, b) is a component of O_1, then

$$f(b, x_2, ..., x_{n-1}, t) - f(a, x_2, ..., x_{n-1}, t) \geq \frac{9}{10} \mu (b - a),$$

where equality holds unless a is an endpoint of J_1 while if (a, b) is a component of O_2, then

$$f(b, x_2, ..., x_{n-1}, t) - f(a, x_2, ..., x_{n-1}, t) \leq - \frac{9}{10} \mu(b - a),$$

with equality unless a is an endpoint of J_1. In the first case since, $\| \frac{\partial f}{\partial x_1} \|_\infty = \mu$, we deduce

$$l \left[(a, b) \cap \{w : \frac{\partial f}{\partial x_1} (w, x_2, ..., x_{n-1}, t) \leq 0\} \right] \leq \frac{1}{10} (b - a),$$

while in the second case, we have

$$l \left[(a, b) \cap \{w : \frac{\partial f}{\partial x_1} (w, x_2, ..., x_{n-1}, t) \geq 0\} \right] \leq \frac{1}{10} (b - a) .$$

If $l(O_1) \geq \frac{2}{3} l(J_1)$, then from the first inequality, we find

$$l(\{z \in J_1 : \frac{\partial f}{\partial x_1} (z, x_2, ..., x_{n-1}, t) > 0\}) > \frac{1}{2} l(J_1),$$

while if $l(O_2) \geq \frac{2}{3} l(J_1)$, we get

$$l(\{z \in J_1 : \frac{\partial f}{\partial x_1} (z, x_2, ..., x_{n-1}, t) < 0\}) > \frac{1}{2} l(J_1).$$

Hence either

(3.7a) $$l[O_1(x_2, ..., x_{n-1}, t)] < \frac{2}{3} l(J_1),$$

or

(3.7b) $$l[O_2(x_2, ..., x_{n-1}, t)] < \frac{2}{3} l(J_1) .$$

Allowing $(x_2, ..., x_{n-1}, t)$ to vary in $\bar{J}_2 \times ... \times \bar{J}_n$, we conclude from the Tonelli theorem and the definition of O_i, $i = 1, 2, ...,$ that there exists a closed set G with

(3.8) $$|G| \geq \frac{1}{6} \mid \bar{Q}_d(x^*, t^*) \mid$$

and either

(3.9a) $$G \subset \bar{Q}_d(x^*, t^*) \cap \{(x,t) : \hat{f}(x,t) = f(x,t)\}$$

or

(3.9b) $$G \subset \bar{Q}_d(x^*, t^*) \cap \{x : \bar{f}(x,t) = f(x,t)\}.$$

Let $\tilde{f} = \hat{f}$ in case (3.9a) holds and $\tilde{f} = \bar{f}$ when (3.9b) is valid. We extend $\tilde{f}$ to R^n in the following way: define $\tilde{f}$ on $R \times \bar{J}_2 \times \ldots \times \bar{J}_n$ by requiring that $\tilde{f}(\cdot, x_2, \ldots, x_{n-1}, t)$ be constant on each component of $R \setminus J_1$ whenever , $(x_2, \ldots, x_{n-1}, t) \in \bar{J}_2 \times \ldots \times \bar{J}_n$.

Next extend $\tilde{f}$ from $R \times \bar{J}_2 \times \ldots \times \bar{J}_n$ to $R \times R \times \ldots \times \bar{J}_n$ by requiring that $\tilde{f}(x_1, \cdot, x_3, \ldots, t)$ be constant on each component of $R \setminus \bar{J}_2$, whenever

$$(x_1, x_3, \ldots, t) \in R \times \bar{J}_3 \times \ldots \times \bar{J}_n .$$

Continuing in this manner, we get $\tilde{f}$ defined on R^n . Now from the definition of $\tilde{f}$ it is easily seen that $\tilde{f}(\cdot, t)$ is Lipschitz on R^{n-1} whenever $t \in R$, with

(3.10a) $$-\mu \le \frac{\partial \tilde{f}}{\partial x_1}(x,t) \le \frac{9}{10}\mu,$$

for almost every $(x,t) \in R^n$ when (3.9a) holds while

(3.10b) $$-\frac{9}{10}\mu \le \frac{\partial \tilde{f}}{\partial x_1}(x,t) \le \mu,$$

for almost every $(x,t) \in R^n$ when (3.9b) is true. Also in either case we have for $n \ge 3$,

(3.11) $$\left\| \frac{\partial \tilde{f}}{\partial x_j} \right\|_\infty \le \left\| \frac{\partial f}{\partial x_j} \right\|_\infty, \; 2 \le j \le n-1,$$

and

(3.12) $$|\tilde{f}(x,s) - \tilde{f}(x,t)| \le c\, a_2 |s-t|^\alpha ,$$

when $(x,s), (x,t) \in R^n$.

Next let $\phi_1 = \tilde{f}$, $\phi_2 = f$. Then from (3.11), (3.12), (1.1)-(1.3), and (1.13), we see that the hypotheses of Lemma 2.1 hold with $\bar{Q}_d(x^*, t^*)$ replaced by $\bar{Q}_{2d}(x^*, t^*)$ and $F = G$. Moreover $b_n = ca_2$,

$$-b_j = e_j = \left\| \frac{\partial f}{\partial x_j} \right\|_\infty, \; 2 \le j \le n-1,$$

and $b_1 = -\mu$, $e_1 = \frac{9}{10}\mu$, when (3.10a) holds , while $b_1 = -\frac{9}{10}\mu$, $e_1 = \mu$, when (3.10b) is true. Choosing

(3.13) $$\gamma = \frac{\mu}{n100c_2(b+e)}$$

in Lemma 2.1, we get a Lipschitz function $\phi : R^n \to R$ which satisfies (2.6)-(2.8)

In the rest of the proof of Lemma 3.1, we assume that case (3.10a) holds. The proof of Lemma 3.1 for case (3.10b) is essentially the same. From (2.6) and (3.13) we see that

$$-\frac{101}{100}\mu \leq \frac{\partial \phi}{\partial x_1}(x,t) \leq \frac{91}{100}\mu, \tag{3.14}$$

$$\left\|\frac{\partial \phi}{\partial x_j}\right\|_\infty \leq \left\|\frac{\partial f}{\partial x_j}\right\|_\infty + \frac{\mu}{100n}, \quad 2 \leq j \leq n-1. \tag{3.15}$$

Let $\psi \in C_0^\infty([-1,1])$, with $|\psi'| \leq 1000$, and $\int_{\boldsymbol{R}} \psi(x)\,dx = 1$. Given $(x_2,\dots,x_{n-1},t) \in R^{n-1}$, and $\epsilon > 0$, let

$$(\phi(\cdot,x_2,\dots,x_{n-1},t) * \psi_\epsilon)(x) = \epsilon^{-1} \int_{\boldsymbol{R}} \phi(z,x_2,\dots,x_{n-1},t)\psi\left(\frac{x-z}{\epsilon}\right) dz$$

for $x \in R$. Let $\alpha_1 = \tan^{-1}(\frac{91}{100}\mu)$, $\alpha_2 = \tan^{-1}(\frac{101}{100}\mu)$, and $\theta = \frac{1}{2}(\alpha_2 - \alpha_1)$, where $0 < \alpha_1, \alpha_2, < \frac{\pi}{2}$. Put

$$\begin{aligned} x_1 &= \cos\theta\,\bar{x}_1 + \sin\theta\,\bar{x}_n, \\ x_j &= \bar{x}_j,\ 1 \leq j \leq n-1, \\ x_n &= -\sin\theta\,\bar{x}_1 + \cos\theta\,\bar{x}_n, \end{aligned} \tag{3.16}$$

and define η_ϵ on R^{n+1} implicitly by

$$\eta_\epsilon(\bar{x}_1,\bar{x}_2,\dots,\bar{x}_n,t) = x_n - (\phi(\cdot,x_2,\dots,x_{n-1},t) * \psi_\epsilon)(x_1),$$

whenever $(\bar{x}_1,\bar{x}_2,\dots,\bar{x}_n,t) \in R^{n+1}$. Using the chain rule and (3.14), (3.16), we deduce

$$\frac{\left|\frac{\partial \eta_\epsilon}{\partial \bar{x}_1}\right|(\bar{x}_1,\dots,\bar{x}_n,t)}{\left|\frac{\partial \eta_\epsilon}{\partial \bar{x}_n}\right|(\bar{x}_1,\dots,\bar{x}_n,t)} = \frac{|\sin\theta + \cos\theta(\frac{\partial \phi}{\partial x_1}(\cdot,x_2,\dots,t) * \psi_\epsilon)(x_1)|}{|\cos\theta - \sin\theta(\frac{\partial \phi}{\partial x_1}(\cdot,x_2,\dots,t) * \psi_\epsilon)(x_1)|} \tag{3.17}$$

$$\leq \tan\left[\frac{\alpha_1+\alpha_2}{2}\right] = \nu \leq \frac{1}{2}\left(\frac{91}{100} + \frac{101}{100}\right)\mu = \frac{24}{25}\mu.$$

Here, we have used the trigometric formula for the tangent of the sum of two angles, and (in the last inequality) the fact that $\tan^{-1} x$ is concave down on $[0,\infty]$. Applying the implicit function theorem, and using (3.17), we obtain that, locally, the equation $\eta_\epsilon(\bar{x}_1,\dots,\bar{x}_n,t) = 0$ defines a function, $\bar{x}_n = \zeta_\epsilon(\bar{x}_1,\dots,\bar{x}_{n-1},t)$, with $\|\frac{\partial \zeta_\epsilon}{\partial \bar{x}_1}(\bar{x}_1,\dots,\bar{x}_{n-1},t)\|_\infty \leq \nu$, for which (3.15) is true with ϕ replaced by ζ_ϵ. Piecing together these local solutions we find from (3.16), (3.17), that there exists a unique function ζ_ϵ on R^{n-1}, satisying for fixed $x_2,\dots,x_{n-1},t$,

$$x_1 = \cos\theta\,\bar{x}_1 + \sin\theta\,\zeta_\epsilon(\bar{x}_1,\dots,\bar{x}_{n-1},t),$$

$$(\phi(\cdot,x_2,\dots,t) * \psi_\epsilon)(x_1) = -\sin\theta\,\bar{x}_1 + \cos\theta\,\zeta_\epsilon(\bar{x}_1,\dots,\bar{x}_{n-1},t).$$

Letting $\epsilon \to 0$, we see there exists $\zeta : R^n \to R$, uniformly Lipschitz in the sense of (1.1), with

$$x_1 = \cos\theta\, \bar{x}_1 + \sin\theta\, \zeta(\bar{x}_1, ..., \bar{x}_{n-1}, t),$$

(3.18)

$$\phi(x_1, ..., x_{n-1}, t) = -\sin\theta\, \bar{x}_1 + \cos\theta\, \zeta(\bar{x}_1, ..., \bar{x}_{n-1}, t),$$

whenever $(\bar{x}, t) \in R^n$ and $(x,t), (\bar{x}, t)$, are related by (3.16) . Moreover,

$$\| \frac{\partial \zeta}{\partial \bar{x}_1} \|_\infty \leq \frac{24}{25}\mu,$$

(3.19)

$$\| \frac{\partial \zeta}{\partial \bar{x}_j} \|_\infty \leq \| \frac{\partial f}{\partial x_j} \|_\infty + \frac{\mu}{100n}, \; 2 \leq j \leq n-1.$$

Next we study the mapping $(x,t) \to (\bar{x}, t)$, for $x \in R^{n-1}$, and $t \in R$. From (3.18) it is easily seen that

$$\bar{x}_1 = \bar{x}_1(x,t) = \cos\theta\, x_1 - \sin\theta\, \phi(x,t),$$

(3.20)

$$\zeta(\bar{x}, t) = \sin\theta\, x_1 + \cos\theta\, \phi(x,t).$$

Differentiating both sides of (3.20) with respect to x_1, squaring the resulting equations, and adding, we get

$$\sqrt{1 + (\frac{\partial \zeta}{\partial \bar{x}_1})^2}\, \frac{\partial \bar{x}_1}{\partial x_1} = \sqrt{1 + (\frac{\partial \phi}{\partial x_1})^2}.$$

Thus,

$$(1 + \mu^2)^{-\frac{1}{2}} \leq \frac{\partial \bar{x}_1}{\partial x_1} \leq (1 + \mu^2)^{\frac{1}{2}}. \tag{3.21}$$

We note from the mean value theorem of calculus that

$$\theta = \tfrac{1}{2}[\tan^{-1}(\frac{101\mu}{100}) - \tan^{-1}(\frac{91\mu}{100})]$$

(3.22)

$$\leq \frac{\mu}{20[1 + (\frac{9}{10}\mu)^2]} \leq \frac{1}{10} \min\{1, \mu^{-1}\}.$$

Given $\bar{x} \in R^{n-1}$, $s, t \in R$, choose $x \in R^{n-1}$ with $x_j = \bar{x}_j, 2 \leq j \leq n-1$, and $\bar{x}_1(x,t) = \bar{x}_1$. Let $\bar{y} = \bar{x}(x,s)$. The existence of x follows easily from (3.20), (3.21), and the fact that ϕ has compact support. Using (3.18)-(3.22), we find that

$$|\zeta(\bar{x},t) - \zeta(\bar{x},s)| \leq |\zeta(\bar{x},t) - \zeta(\bar{y},s)| + |\zeta(\bar{y},s) - \zeta(\bar{x},s)|$$

$$\leq \cos\theta |\phi(x,t) - \phi(x,s)| + \tfrac{24}{25}\mu|\bar{x} - \bar{y}|$$

(3.23)

$$\leq (\cos\theta + \tfrac{24}{25}\mu \sin\theta)|\phi(x,t) - \phi(x,s)|$$

$$\leq 2\,|\phi(x,t) - \phi(x,s)|.$$

Using (3.23), and (2.7) of Lemma 2.1, we deduce first

$$
\begin{aligned}
|\zeta(\bar{x},t) - \zeta(\bar{x},s)| &\leq c \max\{a_1, a_2, a_2^{\frac{1}{\alpha}} \mu^{(1-\frac{1}{\alpha})}\} \, |s - t|^{\alpha}. \\
&= c\, a_2' \, |s - t|^{\alpha}.
\end{aligned} \tag{3.24}
$$

Second, we get

$$
\int_J \frac{(\zeta(\bar{x},t) - \zeta(\bar{x},s))^2}{|s-t|^{1+2\alpha}} \, ds \leq 4 \int_J \frac{(\phi(x,t) - \phi(x,s))^2}{|s-t|^{1+2\alpha}} \, ds \tag{3.25}
$$

when $J \subset R$ is an interval.

From (2.8) of Lemma 2.1 and weak type estimates we see that given $\lambda > 1$ and $x \in R^{n-1}$, there exists, $H = H(\lambda) \subset J_n$ with

$$
\int_{J_n} \frac{[\phi(x,t) - \phi(x,s)]^2}{|s - t|^{1+2\alpha}} \, ds \leq \lambda \, (a_2')^2, \tag{3.26}
$$

for all $t \in H$, where $|J_n \setminus H| \leq c\,\lambda^{-1} |J_n|$. From (3.24) - (3.26) and (3.8), we deduce that there exists c_3, $0 < c_3 < \infty$, and $G_1 \subset G$, closed, with

$$
|G_1| \geq \frac{1}{2} \, |G| \geq \frac{1}{12} \, |Q_d(x^*, t^*)|, \tag{3.27}
$$

and

$$
\int_{\sigma J_n} \frac{(\zeta(\bar{x},s) - \zeta(\bar{x},t))^2}{|s - t|^{1+2\alpha}} \, ds \leq c_3 \log(\sigma) \, (a_2')^2, \tag{3.28}
$$

whenever $\sigma > 2$, $(x,t) \in G_1$, and $(\bar{x},t), (x,t)$ are related by (3.18), (3.20).

Next given $x \in R^{n-1}$, $s,t \in R$ observe from (3.20) that

$$
|\bar{x}_1(x,t) - \bar{x}_1(x,s)| = \sin\theta \, |\phi(x,t) - \phi(x,s)| \leq c\, a_2' \, |s - t|^{\alpha}. \tag{3.29}
$$

Using (3.29), (3.21), and (3.16), we deduce that if $K = \{(\bar{x},t) : (x,t) \subset G_1\}$ and $\hat{x}_i = \bar{x}_i(x^*, t^*)$, $\hat{x} = (\hat{x}_1, ..., \hat{x}_{n-1})$, then K is closed and $K \subset \bar{Q}_{\sigma d}(\hat{x}, t^*)$, provided $\sigma = c_4 (1 + a_2')^{\frac{1}{\alpha}}$, and c_4 is large enough. From (3.19), (3.24), (3.28), we see that Lemma 2.1 may be applied with d replaced by $2\sigma d$, $F = K$, and $\phi_1 = \phi_2 = \zeta$. Also,

$$
\begin{aligned}
-b_1 &= e_1 = \frac{24}{25} \mu, \\
-b_j &= e_j = \|\frac{\partial f}{\partial x_j}\|_{\infty} + \frac{\mu}{100n}, \; 2 \leq j \leq n-1, \\
b_n &= c(\log\sigma)^{\frac{1}{2}} \, a_2'.
\end{aligned}
$$

If $\gamma = \dfrac{\mu}{(b+e)\, c_2 \, 100n}$ in Lemma 2.1, then from the conclusion of this lemma we get f_2, a function from $R^n \to R$, satisfying (3.5) and uniformly Lipschitz in the sense of (1.1). Moreover, from (2.8) of Lemma 2.1, a ballpark estimate, and the equivalence stated in section 1, we see that (1.1) - (1.4) are valid for f_2 with a_2 replaced by $\tilde{a}_2$ and a_1 by ca_1. Also if $E = G_1$, θ is as in Lemma 3.1, and f_1 is defined in terms of f_2 by (3.3), where x, $\bar{x}$, are related by (3.4), then $\phi = f_1 = f$ on G_1 by (3.20), and the fact that $\zeta = f_2$ on K. Next to show that f_1 satisfies (1.1) with a_1 replaced by ca_1, suppose $x,\ y \in R^{n-1}$, with $x_1 \neq y_1$, $x_j = y_j$, $2 \leq j \leq n-1$; $t \in R$. Put $\bar{x} = \bar{x}(x,t)$, $\bar{y} = \bar{y}(y,t)$, and $\tan u = \dfrac{f_2(\bar{y},t) - f_2(\bar{x},t)}{\bar{y}_1 - \bar{x}_1}$, $-\frac{\pi}{2} < u < \frac{\pi}{2}$. Then from (3.3), (3.4), we obtain

$$\frac{|f_1(x,t) - f_1(y,t)|}{|x_1 - y_1|} = \frac{|-\tan\theta + \tan u|}{|1 + \tan u \tan\theta|} = |\tan(u - \theta)|. \tag{3.30}$$

From the definition of θ and (3.5a) we deduce that $|\theta - u| \leq \tan^{-1} 2\mu$. Using (3.30), and the above deduction, we conclude that

$$\left| \frac{f_1(x,t) - f_1(y,t)}{x_1 - y_1} \right| \leq 2\mu \leq 2a_1.$$

Finally, as in (3.23)-(3.24), with the roles of $\bar{x}$, x interchanged we see that f_1 satisfies (1.4) with a_2 replaced by $\tilde{a_2}$. The proof of Lemma 3.1 is now complete. □

4. The David Buildup Scheme : Part 2.

In this section we begin the second part of the David buildup scheme. To this end, let $\tilde{a}_2, \theta, l, f_1, f_2, x, \bar{x}$, be related as in Lemma 3.1. Let $K_{j,\epsilon}$, $H_{j,\epsilon}$, K_j, $\bar{K}_j$, be as in section 1. We write $K_{j,\epsilon,f}$ for $K_{j,\epsilon}$, when we want to indicate the dependence of this operator on f . We first introduce some new operators. Let $\hat{x},\ \hat{y} \in R^{n-1}$, $s,t \in R$, and put

$$L_{1,\epsilon,f_i}(\hat{x},t,\hat{y},s) = \frac{f_i(\hat{x},t) - f_i(\hat{y},s)}{|t-s|^{\frac{n}{2}+1}} \cdot \exp\left[\frac{-|\hat{x}-\hat{y}|^2 - |f_i(\hat{x},t) - f_i(\hat{y},s)|^2}{4|s-t|} \right],$$

when

$$|f_i(\hat{x},t) - f_i(\hat{y},s)|^2 + |\hat{x}-\hat{y}|^2 \geq \epsilon^2,\ i = 1,2,$$

and $|s-t| \geq \epsilon^2$. Put $L_{1,\epsilon,f_i} \equiv 0$, otherwise. Set

$$L_{j,\epsilon,f_i}(\hat{x},t,\hat{y},s) = \frac{\hat{x}_{(j-1)} - \hat{y}_{(j-1)}}{|t-s|^{\frac{n}{2}+1}} \cdot \exp\left[\frac{-|\hat{x}-\hat{y}|^2 - |f_i(\hat{x},t) - f_i(\hat{y},s)|^2}{4|s-t|} \right],$$

when $2 \leq j \leq n$,

$$|f_i(\hat{x},t) - f_i(\hat{y},s)|^2 + |\hat{x}-\hat{y}|^2 \geq \epsilon^2,\ i = 1,2,$$

and $|s-t| \geq \epsilon^2$. Put $L_{j,\epsilon,f_i} \equiv 0$, otherwise. For $n+1 \leq j \leq 2n$ set

$$L_{j,\epsilon,f_i}(\hat{x},t,\hat{y},s) = \operatorname{sgn}(s-t)\, L_{j-n,\epsilon,f_i}(\hat{x},t,\hat{y},s).$$

If $g \in L^p(R^n), 1 < p < \infty$, and $1 \le j \le 2n$ put

$$(4.1) \qquad S_{j,\epsilon,f_i}\, g(\hat{x},t) = \iint_{R^n} L_{j,\epsilon,f_i}(\hat{x},t,\hat{y},s)\, g(\hat{y},s)\, d\hat{y}ds.$$

Using Hölder's inequality it is easily checked that all integrals converge absolutely. We prove

Lemma 4.1 *Fix p, $1 < p < \infty$, and suppose that (1.6) - (1.8) are valid for the operators in Theorem 1 with f replaced by f_2 and c^j by c_j^*, $1 \le j \le n$. Then (1.6) - (1.8) of Theorem 1 also hold for the operators K_{j,ϵ,f_1}, S_{j,ϵ,f_i}, $i = 1,2$; $1 \le j \le 2n$. In fact,*

$$(4.2) \quad \max\{\|\bar{K}_{j,\epsilon,f_1}\, g\|_p\, ,\ \|\bar{S}_{j,\epsilon,f_i}\, g\|_p\, ,\ i = 1,2\} \le c(p)c_j'\, \|g\|_p,\ 1 \le j \le 2n,$$

where

$$c_j' = \begin{cases} (1+a_1)(c_1^* + c_{l+1}^* + \tilde{a}_2), & j \in \{1, n+1, l+1, n+l+1\}, \\ (1+a_1)(c_1^* + 1), & 1 \le j \le 2n,\ j \notin \{1, n+1, l+1, n+l+1\}. \end{cases}$$

Proof. We prove (4.2) for f_1 only when $j = 1$, since the other proofs are essentially the same. Also, we assume, as in section 3, that $l = 1$. To begin the proof, we show that Lemma 4.1 is valid for S_{j,ϵ,f_2}, $j = 1,2$. Indeed, let

(4.3)

$$\begin{aligned} W &= W(\epsilon,\bar{x},t) \\ &= \{(\bar{y},s) : |\bar{x}-\bar{y}|^2 < \epsilon^2 < |\bar{x}-\bar{y}|^2 + |f_2(\bar{y},s) - f_2(\bar{x},t)|^2, \text{ and } |s-t| \ge \epsilon^2\}. \end{aligned}$$

Then, from (1.1), (1.4) for f_2, and (4.3) we deduce

(4.4)

$$\begin{aligned} |(S_{j,\epsilon,f_2} - K_{j,\epsilon,f_2})g(\bar{x},t)| &\le c \iint_W |L_{j,\epsilon,f_2}(\bar{x},t,\bar{y},s)|\, |g(\bar{y},s)|\, d\bar{y}ds \\ &\le c \iint_W (a_1^+\epsilon + a_2^+|s-t|^{\frac{1}{2}})\, |s-t|^{-(\frac{n}{2}+1)}\, |g(\bar{y},s)|\, d\bar{y}ds \\ &\le c \int_{\{s:|s-t|\ge\epsilon^2\}} (a_1^+\epsilon^n + a_2^+\epsilon^{n-1}|s-t|^{\frac{1}{2}})\, |s-t|^{-(\frac{n}{2}+1)}\, M^{(1)}g(\bar{x},s)ds \\ &\le c\,(a_1^+ + a_2^+)\, M^{(2)}(M^{(1)}g)(\bar{x},t) \end{aligned}$$

where $a_1^+ = a_1$, $a_2^+ = \tilde{a}_2$, if $j = 1$ while $a_1^+ = a_2^+ = 1$ for $j = 2$. Also $M^{(1)}$, $M^{(2)}$, denote maximal functions on R^{n-1} , R , with respect to x, t, while the other variables are held constant. The last inequality in (4.4) is proved by writing the integral as a sum over $\{s : 2^k\epsilon^2 \le |s-t| \le 2^{k+1}\epsilon^2\}$ for $k = 0,1,2,\ldots$, and simple estimates. We note that $M^{(2)} \circ M^{(1)}$ can be replaced in (4.4) by a maximal function with respect to 'rectangles' of side length ρ in the space variable and ρ^2 in the time variable. However, to prove this stronger inequality we would need to use the exponential term in L_{j,ϵ,f_2}, and we shall not need such strong estimates until Lemma 4.2. From (4.4) we conclude that

$$\begin{aligned} \sup_{\epsilon>0} | S_{j,\epsilon,f_2}\, g(\bar{x},t)| &= \bar{S}_{j,f_2}\, g(\bar{x},t) \\ &\le \bar{K}_{j,f_2}\, g(\bar{x},t) + c\,(a_1^+ + a_2^+) M^{(2)}(M^{(1)} g)(\bar{x},t). \end{aligned} \tag{4.5}$$

Since (1.8) holds for $\bar{K}_{j,f_2}$ it follows from (4.5) and the Hardy Littlewood Maximal Theorem that (4.2) holds for $\bar{S}_j$ with c_j' replaced by $c_j^* + c(p)(a_1^+ + a_2^+)$.

Also, we claim that

$$\lim_{\epsilon\to 0} S_{j,\epsilon,f_2}\, g(\bar{x},t) = \lim_{\epsilon\to 0} K_{j,\epsilon,f_2}\, g(\bar{x},t)\,. \tag{4.6}$$

To prove (4.6) it sufffices in view of (4.5) , (1.1), (1.4), and a well known argument (see [LM1, section 5]) to prove this equality when $g = 1$. In this case the integrals are interpreted as principal values near ∞. To prove (4.6) when $g = 1$, we note that for almost every $(\bar{y},\tau) \in R^n$,

$$|f_2(\bar{z},\tau) - f_2(\bar{y},\tau) - \langle \nabla_{\bar{y}} f_2(\bar{y},\tau), \bar{z} - \bar{y}\rangle | \to 0, \tag{4.7}$$

as $\bar{z}\to\bar{y}$, and

$$|s-\tau|^{-\frac{1}{2}}\, |f_2(\bar{y},s) - f_2(\bar{y},\tau)| \to 0\,, \tag{4.8}$$

as $s\to\tau$. (4.7) follows from (1.1) while (4.8) is an easy consequence of (1.4) and (1.13) . Using (4.3), (4.7), (4.8), Egoroff's Theorem, and making estimates similar to those in the proof of (1.8) for S_{j,ϵ,f_2}, it can be shown for almost every $(\bar{x},t)$ that

(4.9)

$$\begin{aligned} (S_{1,\epsilon,f_2} - K_{1,\epsilon,f_2})\, 1(\bar{x},t) &= \int\!\!\int_{W(\epsilon,\bar{x},t)} \langle \nabla_{\bar{x}} f_2(\bar{x},t)\,,\, \bar{y}-\bar{x}\rangle\, |s-t|^{-(\frac{n}{2}+1)} \\ &\quad\cdot \exp\left[\frac{-(|\bar{x}-\bar{y}|^2 + \langle \bar{x}-\bar{y},\, \nabla_{\bar{x}} f_2(\bar{x},t)\rangle^2)}{4|s-t|} \right] d\bar{y}ds + o(1) \\ &= I + o(1)\,. \end{aligned}$$

Here $o(1)$ denotes a term which tends to zero with ϵ. Finally it follows again from (4.3), (4.7), (4.8), and symmetry that for almost every $(\bar{x},t)$ with respect to Lebesgue n measure we have

$$I = \int\!\!\int_{W(\epsilon,\bar{x},t)} \dots = \iint_B \dots + o(1) = o(1)\,. \tag{4.10}$$

where $B = \{(\bar{y},s) : |\bar{x}-\bar{y}|^2 < \epsilon^2 < |\bar{x}-\bar{y}|^2 + \langle \bar{x}-\bar{y},\, \nabla_{\bar{x}} f_2(\bar{x},t)\rangle^2$ and $|s-t| \ge \epsilon^2\}$. From (4.9), (4.10), we conclude that (4.6) holds when $j = 1$. Likewise, we find first that (4.6) is true for $j = 2$, and second that Lemma 4.1 is valid for S_{j,ϵ,f_2} , $j = 1, 2$.

Repeating the above argument, we obtain Lemma 4.1 for f_2 and $3 \le j \le 2n$. For a proof similar to the proof used in (4.6) with more details, see section 5 of chapter 1.

Next if $g \in L^p(R^n)$, put

$$\bar{g}(\bar{x},t) = g(x(\bar{x},t),t)\,\frac{\partial x_1}{\partial \bar{x}_1}\,(\bar{x},t),\ (\bar{x},t) \in R^n .$$

We note from (3.3), (3.4), that for $\bar{x} = \bar{x}(x,t)$, $\bar{y} = \bar{y}(y,t)$,

$$(4.11) \qquad |\bar{x}-\bar{y}|^2 + |f_2(\bar{x},t) - f_2(\bar{y},s)|^2 = |x-y|^2 + |f_1(x,t) - f_1(y,s)|^2 .$$

From (4.11), (3.3), (3.4), and the fact that $x(\cdot,t)$ is uniformly Lipschitz, we deduce

$$S_{1,\epsilon,f_1}\ g(x,t) = (-\sin\theta S_{2,\epsilon,f_2} + \cos\theta S_{1,\epsilon,f_2})\bar{g}(\bar{x},t)\,,$$

and so from (4.5) for S_{j,ϵ,f_2}, $j = 1,2$, we find

$$(4.12) \qquad \begin{aligned} ||\bar{S}_{1,\epsilon,f_1}\ g||_p &\le ||\bar{S}_{2,\epsilon,f_2}\ \bar{g}||_p + ||\bar{S}_{1,\epsilon,f_2}\ \bar{g}||_p \\ &\le c(p)\,(\tilde{a_2} + c_1^* + c_2^*)\,||\bar{g}||_p. \end{aligned}$$

To estimate, $||\bar{g}||_p$ we again use (3.3), (3.4) to get, as in (3.21), that for almost every $(x,t) \in R^n$,

$$(4.13) \qquad c^{-1}(1+a_1)^{-1} \le ||\,\frac{\partial x_1}{\partial \bar{x}_1}\,||_\infty \le c(1+a_1)$$

and thereupon that

$$(4.14) \qquad ||\bar{g}||_p \le c(1 + a_1)\,||g||_p.$$

From (4.14) and (4.12) we conclude that (4.2) is true with $\bar{K}_{1,\epsilon,f_1}$ replaced by $\bar{S}_{1,\epsilon,f_1}$. The proof of (4.2) for S_{j,ϵ,f_1}, $2 \le j \le 2n$, is similar. We omit the details.

To prove (4.2) for $\bar{K}_{j,\epsilon,f_1}$ we use Lemma 3.1 and argue as in (4.5) to get

$$(4.15) \qquad \bar{K}_{j,\epsilon,f_1}\ g(x,t) \le \bar{S}_{j,\epsilon,f_1}\ g(x,t) + c(a_1^+ + a_2^+)M^{(2)}(M^{(1)}g)(x,t).$$

From (4.15), we see that (4.2) also holds for $\bar{K}_{j,\epsilon,f_1}$. Finally we show that (1.9) of Theorem 1 holds for K_{j,ϵ,f_1}. Indeed from (4.13) we see that sets of Lebesgue n measure zero are mapped into sets of Lebesgue n measure zero by the transformation $(x,t) \longrightarrow (\bar{x},t)$ and its inverse, given by (3.3), (3.4). Using this fact and (4.6), we obtain that for $1 \le j \le 2n$, $\lim_{\epsilon \longrightarrow 0} S_{j,\epsilon,f_1}\ g(x,t)$ exists for almost every (x,t) in R^n with respect to Lebesgue n measure when $g \in L^p(R^n)$, $1 < p < \infty$. Thus Lemma 4.1 is true for S_{j,ϵ,f_1}, $1 \le j \le 2n$. Finally, to replace S by K in this limit we repeat the argument in (4.7)-(4.11). We omit the details. The proof of Lemma 4.1 is now complete . □

Next in this section we prove a lemma which will be general enough to prove Theorems 1-3 in section 5. In order to avoid confusion in section 5, we let $x = (x_1, ..., x_l)$, be a point in R^l , for the remainder of section 4. Fix α, $0 < \alpha < 1$, and let $Q_d(x^*, t^*)$ be as in (1.11) with n replaced by $l+1$. Let $k \le l$ be a fixed positive integer and suppose that μ is a positive Borel measure on R^{l+1} which satisfies for some θ, $0<\theta\leq\frac{1}{2}$, the condition

$$\theta\, d^{k\alpha+1}\leq\mu(\,\bar{Q}_d(x^*,t^*)\,)\leq\theta^{-1}\,d^{k\alpha+1}, \tag{4.16}$$

whenever (x^*,t^*) is in the support of μ and $d>0$. As in section 1, let $L^p(\mu)$, $1\leq p<\infty$, be equal to the space of Borel measurable functions f on R^{l+1}, defined μ almost everywhere, with $\int_{R^{l+1}}|f|^p\,d\mu<\infty$. Define the maximal function of f, $M_\mu f$, on R^{l+1} by

$$M_\mu f(x^*,t^*)=\sup\left[\mu(Q)^{-1}\int_Q|f|\,d\mu\right].$$

where the sup is taken over all $Q=\bar{Q}_d(x^*,t^*)$ with $\mu(Q_{\frac{d}{2}}(x^*,t^*))\neq 0$. Let $\Psi: R^{l+1}\times R^{l+1}\to R$ be a kernel satisfying for fixed α, as above, and $(x,t),(y,s)\in R^{l+1}$, the condition

$$|\Psi(x,t,y,s)|\leq\gamma_1\,\frac{(|x-y|+h|s-t|^\alpha)^{2m+1}}{|s-t|^{(2m+k+1)\alpha+1}}\,\exp\left[-\frac{|x-y|^{\frac{1}{\alpha}}}{(100)^{\frac{1}{\alpha}}|s-t|}\right]. \tag{4.17}$$

Also if $(x,t),(y,s),(z,\tau)\in R^{l+1}$; $|x-y|\geq 2|x-z|$, and $|s-t|^\alpha\geq 2|t-\tau|^\alpha$, then

(4.18)
$$\begin{aligned}&|\Psi(x,t,y,s)-\Psi(z,\tau,y,s)|+|\Psi(y,s,x,t)-\Psi(y,s,z,\tau)|\\&\leq\gamma_2\,(|x-z|+|t-\tau|^\alpha)\,\frac{(|x-y|+h|s-t|^\alpha)^{2m+1}}{|s-t|^{(2m+k+1)\alpha+1}}\left[\frac{|x-y|^{\frac{1}{\alpha}}}{|s-t|}+1\right]\\&\quad\cdot\left[\frac{1}{|x-y|}+\frac{1}{|s-t|^\alpha}\right]\exp\left[-\frac{|x-y|^{\frac{1}{\alpha}}}{(100)^{\frac{1}{\alpha}}\,|s-t|}\right].\end{aligned}$$

In (4.17), (4.18), $0<\gamma_j<\infty, j=1,2,$ and $h=0$ when $k<l$ while $h=1$ when $k=l$. Put

$$\Psi_\epsilon(x,t,y,s)=\Psi(x,t,y,s),$$

when $|x-y|\geq\epsilon$ and $|s-t|\geq\epsilon^{\frac{1}{\alpha}}$, while otherwise $\Psi_\epsilon(x,t,y,s)=0$. Next set

$$T_{\mu,\epsilon}\,g(x,t)=\int_{R^{l+1}}\Psi_\epsilon(x,t,y,s)\,g(y,s)\,d\mu(y,s)$$

$$\bar{T}_\mu\,g(x,t)=\sup_{\epsilon>0}|\,T_{\mu,\epsilon}\,g(x,t)\,|,\ (x,t)\in R^{l+1}.$$

Then the second part of the David buildup scheme is contained in Lemma 4.2 .

Lemma 4.2. *Let μ be as in (4.16) and define $T_{\mu,\epsilon}$, $\bar{T}_\mu$, as above . Suppose for some fixed $\eta\in(0,\frac{1}{2})$ and all $Q=Q_d(x^*,t^*)\subset R^{l+1}$, with $(x^*,t^*)\in$ supp μ,* *that there exists : (A) a Borel measurable set* $E \subset Q \cap (\text{ supp } \mu)$, *(B) a kernel,* Ψ_Q, *and (C) a locally finite Borel measure* $\nu = \nu_Q$ *on* R^{l+1} , *with the properties*

$$(A) \qquad \mu(E) \geq \eta\, \mu(Q),$$

$$(4.19) \qquad \begin{array}{l} (B)\ \Psi_Q \textit{ satisfies (4.17), (4.18), and } \Psi_Q = \Psi, \textit{ on} \\ (E \times E) \setminus \{(x,t,y,s) : (x,t) = (y,s)\}, \end{array}$$

$$(C)\ \nu \textit{ satisfies (4.16) with } \mu \textit{ replaced by } \nu \textit{ and } E \subset \textit{supp } \nu .$$

Define $T_{\nu,\epsilon,Q}$, $\bar{T}_{\nu,Q}$, *as above with* Ψ, μ, *replaced by* Ψ_Q, ν, *and assume for* $1 < p < \infty$, $g \in L^p(\nu)$, *that*

$$(4.20) \qquad ||\bar{T}_{\nu,Q}\, g||_{p,\nu} \leq c(p)\, \gamma_3\, ||g||_{p,\nu},$$

$$(4.21) \qquad \lim_{\epsilon \to 0} T_{\nu,\epsilon,Q}\, g(x,t) = T_{\nu,Q}\, g(x,t),$$

for almost every $(x,t) \in R^{l+1}$ *where* $0 < \gamma_3 < \infty$, *and* $||\cdot||_{p,\nu}$ *denotes the norm in* $L^p(\nu)$. *Then (4.20), (4.21), are also true for* $\bar{T}_\mu$, $T_{\mu,\epsilon}$, *with* $L^p(\nu)$ *replaced by* $L^p(\mu)$ *and* γ_3 *by* $A_1(\gamma_1 + \gamma_2 + \gamma_3)$, *where* $A_1 = A_1(m, \alpha, \theta, \eta, k, l)$.

Proof. Lemma 4.2 for $\bar{T}_\mu$ is a parabolic version of several theorems of David (see [J, ch 8] [D1]). Since our proof of (4.20) is quite similar to David's proof of Proposition 4 in [D1] , we shall only sketch the details. In fact, we shall often refer to [D1], so the reader is advised to have this paper nearby. We shall need the following covering lemma : Let $\mathcal{G}$ be a covering of $F \subset$ supp μ by 'rectangles' $Q = \bar{Q}_d(x^*, t^*)$ with $(x^*, t^*) \in F$ and $d \leq N < \infty$. Then there exists $c > 2$ and a countable subcovering $\mathcal{G}^*$ of F consisting of 'rectangles' of the form $Q^* = \bar{Q}_{cd}(x^*, t^*)$, with the property that $\{\ \bar{Q}_d(x^*, t^*)\ : Q^* \in \mathcal{G}^*\} \subset \mathcal{G}$ and the members of this set are pairwise disjoint

To begin the proof, we first note that it suffices to prove (4.20) only for compactly supported g in $L^p(\mu)$, $1 < p < \infty$. Indeed, for an arbitrary $g \in L^p(\mu), 1 < p < \infty$, we can choose a sequence (g_j) of functions in $L^p(\mu)$ with compact support which converge pointwise (μ almost everywhere) and in the norm of $L^p(\mu)$ to g. Then from (4.17), and Hölder's inequality, we deduce for fixed $\epsilon > 0$ that $(T_{\mu,\epsilon}\, g_j)$ converges pointwise to $T_{\mu,\epsilon}\, g$ as $j \to \infty$. Thus, $\bar{T}_\mu\, g \leq \liminf_{j \to \infty} \bar{T}_\mu\, g_j$. Using this inequality and the Fatou lemma, we get (4.20) for a general $g \in L^p(\mu)$. Thus we assume that g has compact support. The proof of (4.20) for $\bar{T}_\mu$ consists in establishing the following good λ inequality : given $\epsilon > 0$ there exists,

$$(4.22) \qquad \beta = \frac{\epsilon}{(\gamma_1 + \gamma_2 + \gamma_3)\, A_1}$$

where $A_1 > 2$ is as in Lemma 4.2, such that

$$(4.23) \qquad \begin{aligned} &\mu\left(\{(x,t) \in R^{l+1}\ : \bar{T}_\mu\, g(x,t) > (1+\epsilon)\lambda,\ \hat{g}(x,t) \leq \beta\,\lambda\}\right) \\ &\leq (1 - \tfrac{\eta\theta^2}{c_4})\, \mu\left(\{(x,t) \in R^{l+1}\ : \bar{T}_\mu\, g(x,t) > \lambda\}\right) . \end{aligned}$$

In (4.23), $c_4 \geq 2$ depends only on α and l. Also $\hat{g} = (M_\mu\, |g|^r)^{\frac{1}{r}}$ when $k < l$, while $\hat{g} = [(M^{(2)} \circ M^{(1)} + M^{(1)} \circ M^{(2)})(\chi |g|^r)]^{\frac{1}{r}}$ when $k = l$, where $M^{(1)}$, $M^{(2)}$ are defined as in (4.4), $r = \dfrac{1+p}{2}$, and χ is the characteristic function of the support of μ. We observe that the right hand side of (4.23) is finite for each $\lambda > 0$, as we deduce from (4.16) - (4.18), and our assumption that g has compact support. Using this fact and applying the above covering lemma, we get a covering $\{Q_{cd_i}(x_i, t_i)\}$ of

$$F = (\operatorname{supp} \mu) \cap \{(x,t) : \bar{T}_\mu\, g(x,t) > \lambda\},$$

with $(x_i, t_i) \in F$ such that

$$(a)\ \cup Q_{d_i}(x_i, t_i) \subset \{(x,t) : \bar{T}_\mu\, g(x,t) > \lambda\},$$

$$(b)\ Q_{cd_i}(x_i, t_i) \cap \{(x,t) : \bar{T}_\mu\, g(x,t) \le \lambda\} \neq \emptyset,$$

$$(c)\ Q_{d_i}(x_i, t_i) \text{ are pairwise disjoint .}$$

From (c) and (4.16) we deduce that

$$\theta^2\, \mu(F) \le c\, \mu(\cup Q_{d_i}(x_i, t_i)).$$

From this inequality and (a) it is easily seen that it suffices to prove (4.23) with θ^2 replaced by 1 and R^{l+1} by $Q = Q_d(x^*, t^*)$, where $Q_d(x^*, t^*) \subset \{(x,t) : \bar{T}_\mu\, g(x,t) > \lambda\}$ and $(x^*, t^*) \in \operatorname{supp} \mu$. Moreover from (b) we can also assume that there exists (z, τ) and $c_5 \ge 2$ with $|z - x^*| + |\tau - t^*|^\alpha \le (c_5\, d)^\alpha$ and

$$|\bar{T}_\mu\, g(z, \tau)| \le \lambda\,. \tag{4.24}$$

Let E be as in 4.19 (A) , defined relative to $Q' = Q_{\frac{d}{2}}(x^*, t^*)$, and choose E_1 a closed subset of E with

$$\mu(E_1)/\mu(Q') \ge \frac{\eta}{2}\,. \tag{4.25}$$

In the rest of the proof of Lemma 4.2, we let A denote a positive constant depending only on $k, l, \alpha, m, \theta, \eta$, not necessarily the same at each occurrence. Moreover, as usual, c denotes a positive constant depending only on l and α. For $(x,t) \in Q' \cap \operatorname{supp} \mu$ we claim that

$$|\bar{T}_\mu(g\, \chi_{R^{l+1} \setminus Q})(x,t)| \le \lambda + A(\gamma_1 + \gamma_2) \inf_{(y,s) \in Q} \hat{g}(y, s)\,. \tag{4.26}$$

To prove (4.26) let $\hat{Q} = Q_{4c_5 d}(x^*, t^*)$ and write $R^{l+1} \setminus \hat{Q} = \cup_{i=1}^{3} F_i$, where

$$F_1 = \{(y,s) : |y - x^*| \le (2c_5 d)^\alpha\} \cap (R^{l+1} \setminus \hat{Q}),$$

$$F_2 = \{(y,s) : |s - t^*| \le 2c_5 d\} \cap (R^{l+1} \setminus \hat{Q}),$$

$$F_3 = \{(y,s) : |s - t^*| > 2c_5 d \text{ and } |y - x^*| > (2c_5 d)^\alpha\} \cap (R^{l+1} \setminus \hat{Q}).$$

Then from (4.24) and (4.17), we deduce for $(x,t) \in Q' \cap$ supp μ that

(4.27)

$$
\begin{aligned}
|\bar{T}_\mu(g\,\chi_{R^{l+1}\setminus Q})(x,t)| \leq\ & \lambda + A\gamma_1 \inf_{(y,s)\in Q} \hat{g}(y,s) \\
&+ A \int_{F_1} |\Psi(x,t,y,s) - \Psi(z,\tau,y,s)|\, |g|(y,s) d\mu(y,s) \\
&+ A \int_{F_2} |\Psi(x,t,y,s) - \Psi(z,\tau,y,s)|\, |g|(y,s)\, d\mu(y,s) \\
&+ A \int_{F_3} |\Psi(x,t,y,s) - \Psi(z,\tau,y,s)|\, |g|(y,s)\, d\mu(y,s) \\
= \lambda + A\gamma_1 \inf_{(y,s)\in Q} \hat{g}(y,s) &+ I_1 + I_2 + I_3.
\end{aligned}
$$

Again from (4.17), our choice of (z,τ), and the fact that $(x,t) \in Q'$, we see for $j = 1, 2$, that

(4.28)

$$
I_j \leq A\gamma_1 \int_{F_j} \frac{(|x^* - y| + h|s - t^*|^\alpha)^{2m+1}}{|s - t^*|^{(2m+k+1)\alpha+1}} \exp\left[- \frac{|x^* - y|^{\frac{1}{\alpha}}}{c\,|s - t^*|} \right] |g|(y,s) d\mu(y,s)\,.
$$

If $j = 1$ in (4.28) we consider two cases. First if $k < l$, then $h = 0$ in (4.28), and we put

$$
H_i = \{(y,s) : 2^i\, c_5\, d \leq |s - t^*| \leq 2^{i+1}\, c_5\, d\} \cap \{(y,s) : |y - x^*|^{\frac{1}{\alpha}} \leq 2^{i+1}\, d\},
$$

for $i = 1, 2, \ldots$, and get from (4.16), (4.28) , that for $(x,t) \in Q' \cap$ supp μ,

$$
\begin{aligned}
I_1 &\leq A\gamma_1\, d^{(2m+1)\alpha} \sum_{i=1}^{\infty} \int_{H_i} |s - t^*|^{-(1+[2m+1+k]\alpha)}\, |g|(y,s) d\mu(y,s) \\
&\leq A\gamma_1 \Big(\sum_{i=1}^{\infty} 2^{-i(2m+1)\alpha}\Big) \inf_{(y,s)\in Q} M_\mu\, g(y,s) \qquad (4.29) \\
&\leq A\gamma_1 \inf_{(y,s)\in Q} \hat{g}(y,s)\ ,
\end{aligned}
$$

where we have used Hölder's inequality to get the last inequality. If $k = l$, then $h = 1$ in (4.17) and the kernel in (4.17) has a larger growth as $s \to \infty$, than when $h = 0$. To handle this case we note from (4.16) that μ is absolutely continuous with respect to Lebesgue measure on R^{l+1} , and on supp μ we have

$$
\theta\, dxdt \leq d\mu \leq \theta^{-1}\, dxdt.
$$

Using this fact in (4.28), we find that

$$I_1 \le A\gamma_1 d^{1+l\alpha} \int_{\{s:|s-t^*|\ge 2c_6 d\}} |s-t^*|^{-(1+l\alpha)} \inf_{\{y:|y-x^*|<(2c_6d)^\alpha\}} M^{(1)} g(y,s)ds$$

$$\le A\gamma_1 \inf_{(y,s)\in Q} \hat{g}(y,s),$$

thanks again to Hölder's inequality . Thus in either case (4.29) is true.

If $j = 2$ in (4.28) we put

$$\tilde{H}_i = \{(y,s) : |s-t^*| \le 2^{i+1} c_5 d\} \cap \{(y,s) : 2^i c_5 d \le |y-x^*|^{\frac{1}{\alpha}} \le 2^{i+1} c_5 d\},$$

when $i = 1, 2, ...$ and use the fact that for $q > 0$, we have $v_2^q e^{-v_2} \le c(q) v_1^q e^{-v_1}$ whenever $1 \le v_1 \le v_2 < \infty$. We get for $(x,t) \in Q'$,

(4.30)

$$I_2 \le A\gamma_1 d^{-(1+[2m+k+1]\alpha)} \int_{F_2} |x^*-y|^{2m+1} \exp\left[-\frac{|x^*-y|^{\frac{1}{\alpha}}}{c\,d}\right] |g|(y,s)d\mu(y,s)$$

$$\le A\gamma_1 d^{-(1+k\alpha)} \left\{\sum_{i=1}^{\infty} 2^{i(2m+1)\alpha} \exp\left[-\tfrac{2^i}{c}\right] \int_{\tilde{H}_i} |g|(y,s)\, d\mu(y,s)\right\}$$

$$\le A\gamma_1 \inf_{(y,s)\in Q} \hat{g}(y,s) \left\{\sum_{i=1}^{\infty} 2^{(i[2m+1+k]\alpha+1)} \exp\left[-\tfrac{2^i}{c}\right]\right\}$$

$$\le A\gamma_1 \inf_{(y,s)\in Q} \hat{g}(y,s) .$$

To estimate I_3, let

$$D_1 = \{(y,s)\in F_3 : |y-x^*| \le |s-t^*|^\alpha\},$$

$$D_2 = \{(y,s)\in F_3 : |y-x^*| \ge |s-t^*|^\alpha\} .$$

Then from (4.18) we see that

$$I_3$$

$$\leq A\gamma_2 d^\alpha \sum_{i=1}^{2} \int_{D_i} \frac{(|x^* - y| + h|s - t^*|^\alpha)^{2m+1}}{|s - t^*|^{(2m+k+1)\alpha+1}} \left[\frac{1}{|x^* - y|} + \frac{1}{|s - t^*|^\alpha} \right]$$

$$\cdot \left[\frac{|x^* - y|^{\frac{1}{\alpha}}}{|s - t^*|} + 1 \right] \exp\left[- \frac{|x^* - y|^{\frac{1}{\alpha}}}{c|s - t^*|} \right] |g|(y, s)\, d\mu(y, s)$$

$$= J_1 + J_2,$$

where J_1, J_2, are the first and second terms in the above sum . We write the integral in J_1, J_2, as a sum over, $\{s : 2^i c_5 d \leq |s - t^*| \leq 2^{i+1} c_5 d\}$, for $i = 1, 2, \ldots$. We then estimate each term of the resuting sum in J_1, J_2, as in (4.29), (4.30), respectively, with d replaced by $|s - t^*|$. We obtain

$$I_3 \leq A\,\gamma_2 \inf_{(y,s)\in Q} \hat{g}(y, s)\ . \tag{4.31}$$

Using (4.29) - (4.31) in (4.27) we find that claim (4.26) is true.

Now suppose that $\Psi_{Q'}$, $T_{\nu,Q'}$, $T_{\nu,\epsilon,Q'}$ are as in Lemma 4.2. To continue with the proof of (4.23), we note that if $\hat{g} > \beta\,\lambda$ on Q, then (4.23) is trivially true. Hence we assume that

$$\inf_{(y,s)\in Q} \hat{g}(y, s) \leq \beta\lambda\,. \tag{4.32}$$

Let $g_1 = g\,\chi_Q$. Then in view of (4.26), (4.19)(A), and (4.32), we deduce that in order to prove (4.23) it suffices to show

$$\mu(E_1 \cap \{(y, s) \in Q' : \bar{T}_\mu\, g_1(y, s) \leq \frac{\epsilon\lambda}{2} \}) \geq \frac{\eta}{4}\ \mu(Q'), \tag{4.33}$$

provided A_1 is suitably large . To obtain (4.33), we observe that

$$T_{\mu,\epsilon}\, g_1 = (T_{\mu,\epsilon} - T_{\mu,\epsilon,Q'})\, g_1 + T_{\mu,\epsilon,Q'}\, g_1\,.$$

Thus,

$$\mu(\{(y, s) \in E_1 : \bar{T}_\mu\, g_1(y, s) > \frac{\epsilon\lambda}{2} \})$$

$$\leq \mu(\{(y, s) \in E_1 : \overline{T_\mu - T_{\mu,Q'}}\, g_1(y, s) > \frac{\epsilon\lambda}{4} \})$$

$$+ \mu(\{(y, s) \in E_1 : \bar{T}_{\mu,Q'}\, g_1(y, s) > \frac{\epsilon\lambda}{4} \}) \tag{4.34}$$

$$= \sigma_1 + \sigma_2\,.$$

Now,

(4.35)
$$\frac{\epsilon\lambda}{4}\,\sigma_1 \le \int_{E_1} |\overline{T_\mu - T_{\mu,Q'}}\, g_1|(x,t)\, d\mu(x,t)$$

$$\le \int_{Q\setminus E_1} \left(\int_{E_1} |\Psi(x,t,y,s) - \Psi_{Q'}(x,t,y,s)\,|\, d\mu(x,t) \right) |g(y,s)|\, d\mu(y,s)$$

$$= \sigma_3\,,$$

where we have used (4.19)(B) to get the last inequality . To estimate σ_3, fix $(y,s) \in Q \setminus E_1$ and let $\rho = d(\{(y,s)\}, E_1)$ be the nonisotropic distance from (y,s) to E_1 . We write $E_1 = L_1 \cup L_2$, where

$$L_2 = \{(x,t) \in E_1 : |y-x| > 2\rho,\ |s-t|^\alpha > 2\rho\},$$

$$L_1 = E_1 \setminus L_2\,.$$

Then from (4.17) we see as in (4.29) and (4.30) that for $(y,s) \in Q \setminus E_1$,

$$\int_{L_1} |\Psi(x,t,y,s) - \Psi_{Q'}(x,t,y,s)| d\mu(x,t)$$

(4.36)
$$\le \int_{L_1} (|\Psi(x,t,y,s)| + |\Psi_{Q'}(x,t,y,s)|)\, d\mu(x,t) \ \le A\gamma_1\,.$$

To handle the integral over L_2, choose $(w,u) \in E_1$ with $|y-w| + |s-u|^\alpha = \rho$. Then from (4.19)(B), we have $\Psi_{Q'}(x,t,w,u) = \Psi(x,t,w,u)$, when $(x,t) \in E$ and $(x,t) \ne (w,u)$. Thus for $(x,t) \in L_2$,

$$|\Psi(x,t,y,s) - \Psi_{Q'}(x,t,y,s)|$$

$$\le |\Psi(x,t,y,s) - \Psi(x,t,w,u)| + |\Psi_{Q'}(x,t,w,u) - \Psi_{Q'}(x,t,y,s)|\,.$$

Using this inequality, (4.18), and arguing as in (4.31) we find

(4.37)
$$\int_{L_2} |\Psi(x,t,y,s) - \Psi_{Q'}(x,t,y,s)|\, d\mu(x,t) \le A\gamma_2\,.$$

From (4.35) - (4.37) and (4.32), we conclude that

$$\sigma_1 \le \frac{4}{\epsilon\lambda}\,\sigma_3 \ \le \frac{4}{\epsilon\lambda}\, A(\gamma_1 + \gamma_2)\,\mu(Q') \inf_{(y,s)\in Q} \hat{g}(y,s)$$

(4.38)
$$\le \frac{\eta}{8}\,\mu(Q'),$$

for A_1 sufficiently large.

To estimate σ_2, let $q = \frac{p}{p-1}$, and let $T^*_{\nu,\epsilon,Q'}$ be the adjoint operator of $T_{\nu,\epsilon,Q'}$, as an operator on $L^p(\nu)$. That is,

$$T^*_{\nu,\epsilon,Q'}\, h(x,t) \;=\; \int_{R^{l+1}} \Psi_{Q',\epsilon}(y,s,x,t)\, h(y,s)\, d\nu(y,s)\,,$$

when $h \in L^q(\nu)$. From duality, and (4.20), we see that $T^*_{\nu,\epsilon,Q'}$ maps $L^q(\nu)$ into itself with the same constants as for $\bar{T}_{\nu,Q'}$ in (4.20). David (see [D1, Proposition 2]) in fact shows that if $1 < q < \infty$ and $h_1 \in L^q(\nu)$, then

$$(4.39)\qquad ||T^*_{\nu,\epsilon,Q'}\, h_1||_{q,\mu} \;\le\; c(q)\, A\,(\gamma_1 \,+\, \gamma_2 \,+\, \gamma_3\,)||h_1||_{q,\nu}\,.$$

To prove (4.39) he first uses estimates similar to those in the proof of (4.26) to obtain the following generalization of an inequality of Cotlar (see [D1 , Lemma 3])

$$\tilde{T}^*_{\nu,\epsilon,Q'} h_1 \;\le\; c\, M_\nu(T^*_{\nu,\epsilon,Q'}\, h_1) \,+\, A\,(\gamma_1 \,+\, \gamma_2)\,\tilde{h}_1\,,$$

where $\tilde{h}_1$ is obtained by replacing p by q, and μ by ν in the definition of $\hat{h}_1$. (4.39) follows from this inequality and the fact that M_ν is bounded from $L^s(\nu)$ into $L^s(\mu)$ by A, when $1 < s < \infty$. This fact can be proved by the usual argument from the above covering lemma (see [S, ch 1]). From (4.39), (4.17), (4.18), and Hölder's inequality, we observe that if h_1, h_2, have compact support in R^{l+1} , $h_1 \in L^q(\nu)$, and $h_2 \in L^p(\mu)$, then

$$\int_{R^{l+1}} h_1(x,t)(T_{\mu,\epsilon,Q'}\, h_2)(x,t)d\nu(x,t) = \int_{R^{l+1}} (T^*_{\nu,\epsilon,Q'}\, h_1\,)(y,s)\, h_2(y,s)d\mu(y,s)$$

$$\le\; ||\, T^*_{\nu,\epsilon,Q'}\, h_1\,||_{q,\mu}\, ||h_2||_{p,\mu} \le\, c(q)\, A\,(\gamma_1 \,+\, \gamma_2 \,+\, \gamma_3\,)\,||h_1||_{q,\nu}\, ||h_2||_{p,\mu}.$$

From this inequality, we conclude, first, that $\{T_{\mu,\epsilon,Q'}\}$ is uniformly bounded from $L^p(\mu)$ into $L^p(\nu)$; and second, that a subsequence, say $\{T_{\mu,\epsilon_j,Q'}\}$, tends weakly as $\epsilon_j \to 0$, to $T_{\mu,Q'}$ with

$$(4.40)\qquad ||T_{\mu,Q'}\, h_2\,||_{p,\nu} \;\le\; c(p)\, A\,(\gamma_1 \,+\, \gamma_2 \,+\, \gamma_3)\;||h_2||_{p,\mu}.$$

Next we again use an argument of Cotlar ([D1, Lemma 5]) to deduce from (4.40) , (4.17), and (4.18) for $(x,t) \in E_1$ that

$$(4.41)\quad \bar{T}_{\mu,Q'}\, h_2(x,t) \;\le\; c\, M_\nu\, (T_{\mu,Q'}\, h_2)(x,t) \,+\, A\,(\gamma_1 + \gamma_2 \,+\, \gamma_3)\, \hat{h}_2(x,t)\,.$$

We note from (4.16), (4.19)(C), and basic differentiation theory that there exists f, a Borel measurable function defined on E_1, with

$$(4.42)\qquad \mu(H) \;=\; \int_H f\, d\nu\,, H \subset E_1,\; H \;\;\text{Borel},$$

and $|f| \;\le\; \theta^{-2}$. We also note as above that $\hat{h}_2$, M_ν are bounded on $L^p(\mu)$, $L^p(\nu)$ respectively, by A. From these notes, (4.40) - (4.42) with $h_2 = g_1$, and the usual weak type estimates , we see that

$$(\tfrac{\epsilon\lambda}{4})^r\,\sigma_2 \le \int_{E_1} |\bar{T}_{\mu,Q'}\,g_1(x,t)|^r\,d\mu(x,t)$$

$$\le A^r\,||M_\nu T_{\mu,Q'}\,g_1||^r_{r,\nu} + A^r\,(\gamma_1 + \gamma_2 + \gamma_3)^r\,||\hat{g}_1||^r_{r,\mu} \tag{4.43}$$

$$\le A^r\,(\gamma_1 + \gamma_2 + \gamma_3)^r||g_1||^r_r$$

$$\le A^r\,(\gamma_1 + \gamma_2 + \gamma_3)^r\,[\inf_{(y,s)\in Q} \hat{g}(y,s)\,]^r\,\mu(Q')\,.$$

Multiplying (4.43) by $(\frac{4}{\epsilon\lambda})^r$, we conclude from (4.32) and (4.22) that

$$\sigma_2 \le \frac{\eta}{8}\,\mu(Q'), \tag{4.44}$$

for A_1 large enough. Using (4.44) and (4.38) in (4.34), we conclude first that (4.33) is true and thereupon that (4.23) is valid. To prove (4.20), we multiply both sides of (4.23) by λ^{p-1} and integrate. If $\epsilon = \frac{\eta\theta^2}{cp}$ and c is large, we obtain (see [D1, Lemma 7])

$$||\bar{T}_\mu\,g||_{p,\mu} \le c(p)\,A\,(\gamma_1 + \gamma_2 + \gamma_3\,)||g||_{p,\mu}\,,$$

which is (4.20) for $\bar{T}_\mu$.

The proof of (4.21) for T_μ is by contradiction. From (4.17), (4.18), (4.20) for $\bar{T}_\mu$, and a standard Calderon-Zygmund type argument, we see that it suffices to prove (4.21) for $T_{\mu,\epsilon}$ when $g = \chi_{Q^+}$, where Q^+ is a closed ' rectangle ' in R^n . Given r, $0 < r < \frac{1}{2}$, let

$$G = G(r) = \{(x,t) \in R^n\ :[\limsup_{\epsilon\to 0} T_{\mu,\epsilon}(\chi_{Q^+}) - \liminf_{\epsilon\to 0} T_{\mu,\epsilon}(\chi_{Q^+})](x,t) > r\}.$$

From (4.17), (4.18), we see that $G \subset Q^+$. If G has positive μ measure, then since every point of G (except on a set of μ measure 0) has μ density 1, we can find an open 'rectangle' $\hat{Q} \subset Q^+$, such that

$$\mu(\hat{Q}\cap G) > (1 - \frac{\eta}{2})\,\mu(\hat{Q})\,.$$

From this inequality and (4.19) with E defined relative to $\hat{Q}$ we get G_1 closed $\subset \hat{Q}\cap G$, with $\mu(G_1) \ge \frac{\eta}{2}\mu(\hat{Q})$, and $\Psi_{\hat{Q}}$, ν, with $\Psi_{\hat{Q}}\,(x,t,y,s) = \Psi\,(x,t,y,s)$ whenever $(x,t),(y,s) \in G_1, (x,t) \ne (y,s)$. Also, ν satisfies (4.16), $G_1 \subset$ supp ν, and (4.20), (4.21) hold for $T_{\nu,\epsilon}$. Again from differentiation theory, we can choose $Q = Q_{2d}(x^*,t^*) \subset \hat{Q}$ with

$$\mu(Q\cap G_1) \ge (1 - \frac{1}{A_2})\,\mu(Q)\,, \tag{4.45}$$

where

$$A_2 = A_2(m,\alpha,k,l,p,r,\theta,\gamma_1,\gamma_2,\gamma_3) > 2\,,$$

will be chosen later. Put $Q' = \bar{Q}_d(x^*, t^*)$, and set

$$V_\epsilon^{(1)} = T_{\mu,\epsilon}(\chi_Q) - T_{\mu,\epsilon,\hat{Q}}(\chi_Q),$$

$$V_\epsilon^{(2)} = T_{\mu,\epsilon,\hat{Q}}(\chi_{Q\setminus G_1}),$$

$$V_\epsilon^{(3)} = T_{\mu,\epsilon,\hat{Q}}(\chi_{G_1\cap Q})$$

$$\bar{V}^{(j)}(x,t) = \sup_{\epsilon>0}|V_\epsilon^{(j)}|(x,t), j = 1, 2.$$

We observe from (4.42) and (4.21) that

$$\lim_{\epsilon\longrightarrow 0} V_\epsilon^{(3)}(x,t) = \lim_{\epsilon\longrightarrow 0} T_{\nu,\epsilon,\hat{Q}}(f\,\chi_{G_1\cap Q})(x,t)$$

exists for μ almost every $(x,t) \in G_1 \cap Q'$. From this equality, the fact that $T_{\mu,\epsilon,\hat{Q}}(\chi_Q) = \sum_{j=1}^{3} V_\epsilon^{(j)}$ and (4.17) - (4.18) , we deduce that

(4.46)

$$G_1 \cap Q' = \{(x,t) \in Q' : (\limsup_{\epsilon\longrightarrow 0} T_{\mu,\epsilon} - \liminf_{\epsilon\longrightarrow 0} T_{\mu,\epsilon})(\chi_Q)(x,t) > r\}$$

$$\subset \cup_{j=1}^{2} \{(x,t) \in Q' : \bar{V}^{(j)}(x,t) > r/4\}.$$

Now using (4.19)(B), (4.45), and arguing as in (4.35) - (4.37), with $E_1 = Q' \cap G_1$, we get

(4.47)

$$\int_{Q'\cap G_1} |\bar{V}^{(1)}(x,t)|\,d\mu(x,t) \leq A(\gamma_1 + \gamma_2)\,\mu(Q \setminus G_1)$$

$$\leq \frac{(\gamma_1 + \gamma_2)A}{A_2}\,\mu(Q).$$

(4.47), (4.16), and weak type estimates give

(4.48)

$$\mu(\{(x,t) \in Q' \cap G_1 : \bar{V}^{(1)}(x,t) > \frac{r}{4}\}) \leq (\frac{cA(\gamma_1 + \gamma_2)}{A_2 r})\mu(Q) \leq \frac{1}{4}\mu(Q'),$$

for A_2 large enough. Next as in (4.43) - (4.44) , with $g_1 = \chi_{Q\setminus G_1}$, we have

(4.49)

$$\mu(\{(x,t) \in Q' \cap G_1 : \bar{V}^{(2)}(x,t) > \tfrac{r}{4}\})$$

$$\leq A^r\,(\gamma_1 + \gamma_2 + \gamma_3)^r\, r^{-\frac{p+1}{2}}\,\mu(Q \setminus G_1) \leq \frac{1}{4}\,\mu(Q')$$

for A_2 large enough, thanks to (4.45) . Using (4.46), (4.48), and (4.49), we conclude that

$$\mu(Q' \cap G_1) \leq \frac{1}{2}\,\mu(Q'),$$

which in view of (4.16) contradicts (4.45), for A_2 large enough. Hence (4.21) is true for T_μ. The proof of Lemma 4.2 is now complete. □

5. Proof of Theorems 1-3.

Proof. We first prove Theorem 1. Recall from our work in section 1 that if a_1 is sufficiently small ($a_1 \le a_0$), then Theorem 1 is true. Hence Theorem 1 is valid when

$$|||\nabla_x f||| = \sum_{i=1}^{n-1} \| \frac{\partial f}{\partial x_i} \|_\infty \le a_0 . \tag{5.1}$$

Suppose, by way of induction, that Theorem 1 is valid whenever f satisfies (1.2), (1.3), for $0 < a_2 < \infty$, and is uniformly Lipschitz on R^{n-1} in some orthogonal coordinate system, $x_1, ..., x_{n-1}$, with

$$|||\nabla_x f||| \le a . \tag{5.2}$$

Now suppose that f satisfies (1.2), (1.3), for some a_2, $0 < a_2 < \infty$, and is uniformly Lipschitz on R^{n-1} with

$$a_0 \le |||\nabla_x f||| \le (1 - \frac{1}{50(n-1)})^{-1} a , \tag{5.3}$$

in an orthogonal coordinate system. Then given $d > 0$ and $(x^*, t^*) \in R^n$, we see from Lemma 3.1 and (3.6) that there exists f_2, satisfying the hypotheses of Theorem 1, with

$$|||\nabla_x f_2||| \le a, \tag{5.4}$$

in a certain orthogonal coordinate system. Moreover, there exists f_1 related to f_2 by (3.3) with $f_1 = f$ on $E \subset \bar{Q}_d(x^*, t^*) = Q$, where

$$|E| \ge \frac{1}{12} |Q| . \tag{5.5}$$

From (5.4) and the induction hypothesis (5.2), we see that Theorem 1 holds for f_2. Applying Lemma 4.1, we get Theorem 1 for f_1. Finally, from Lemma 3.1 and (1.1), (1.4), we deduce that (4.17)-(4.18) are valid with $\Psi_\epsilon = K_{j,\epsilon,f}$ and $\Psi_{Q,\epsilon} = K_{j,\epsilon,f_1}$, $1 \le j \le 2n$. Since Q is arbitrary and (5.5) is true, we can apply Lemma 4.2 with $k = l = n - 1$ and μ equal to Lebesgue measure on R^n. We conclude from this lemma that (1.6) - (1.8) of Theorem 1 are valid for $\bar{K}_{j,f}$, $K_{j,f}$ when (5.3) holds. Hence by induction, we obtain Theorem 1. □

Proof. To obtain Corollary 1, we first note that since Theorem 1 is true, it follows from (4.2) with $f = f_2$ that (1.6) - (1.8) also hold for $S_{j,\epsilon}$, $1 \le j \le n$, defined relative to f. Since

$$4(4\pi)^{\frac{n}{2}} \mathcal{L}_\epsilon g = (S_{n+1,\epsilon} - S_{1,\epsilon}) g - \sum_{j=2}^{n} (S_{j,\epsilon} - S_{n+j,\epsilon}) (\frac{\partial f}{\partial y_j} g),$$

and (1.1) is true, Corollary 1 follows from Theorem 1 . □

Proof. Next we prove Theorem 2. Given, $d > 0$, and $(x^*, t^*) \in R^n$, let $Q = \bar{Q}_d(x^*, t^*)$, be as in the proof of Theorem 1. Let $F = E_1$, where E_1 is as in (1.16), and set

$$\begin{aligned} \phi_1 &= \phi_2 = f, \\ -b_i &= e_i = \max\{a_1, \delta^{-\frac{1}{2}} a_2\}, \ 1 \le i \le n-1, \\ b_n &= \delta^{-\frac{1}{2}} a_2, \\ \gamma &= 1/c_2 . \end{aligned} \tag{5.6}$$

From (1.15), (1.4), and (1.1), we see that Lemma 3.1 may be applied to get a function ϕ on R^n satisfying (1.1) - (1.4) with constants a_1^*, a_2^*, where

$$\max\{a_1^*, a_2^*\} \le c \max\{a_1, \delta^{-\frac{1}{2}} a_2\} . \tag{5.7}$$

Moreover, $\phi = f$ on E_1. Now we can apply either Theorem 1 of chapter 1 or Theorem 1 of this chapter to the corresponding operators involving ϕ. Using this fact, arbitrariness of Q, and (1.16), we conclude from Lemmas 4.1 and 4.2 with k, l, μ as in the proof of Theorem 1, that (1.6)-(1.7) are valid. Hence, Theorem 2 is true. □

Proof. To begin the proof of Theorem 3, let f be as in the statement of this theorem and recall that $\zeta(x,t) = (x, f(x,t), t)$, when $(x,t) \in R^n$, while $\mathcal{S} = \{\zeta(x,t) : (x,t) \in R^n\}$. Also, as in section 1, let

$$\sigma(F) = \int_{\zeta^{-1}(F \cap \mathcal{S})} \sqrt{1 + |\nabla_y f(y,s)|^2} \, dyds,$$

when F is a Borel subset of R^{n+1}. As in section 1 we denote points in R^{n+1} by (X,t), where $X = (x_1, ..., x_n) = (x, x_n)$ and let $Q_d(X^*, t^*)$ denote the $n+1$ dimensional cube defined as in
(1.11) with n replaced by $n+1$ and (x^*, t^*) by (X^*, t^*). We shall first show that σ satisfies (4.16) with $k = n-1$, $l = n$, and $\theta^{-1} = c(1 + a_1 + a_2)^{4n+4}$, provided c is suitably large. That is, for this value of θ,

$$\theta \rho^{\frac{n+1}{2}} \le \sigma(Q_\rho(X,t)) \le \theta^{-1} \rho^{\frac{n+1}{2}} \tag{5.8}$$

whenever $(X,t) \in \mathcal{S} = \operatorname{supp} \sigma$ and $\rho > 0$. To do this for given $\rho > 0$, let $(X_1, t_1) = (x_1, f(x_1, t_1), t_1) \in \mathcal{S}$, and put $G = \zeta^{-1}(Q_\rho(X_1, t_1))$. Given $r, d > 0$, and $(x^*, t^*) \in R^n$, put $Q^* = Q_r(x_1, t_1)$ and suppose that $Q = \bar{Q}_d(x^*, t^*) \subset Q_{2r}(x_1, t_1)$. Let $m_{j,Q} = m_{j,Q,f}$, $1 \le j \le n-1$, and $m_Q = m_{Q,f}$ denote the average of $\dfrac{\partial f}{\partial x_j}$, f respectively over Q as defined in section 1. Also, put

$$I = J_1 \times \dots \times J_{n-1} = \{x : |x_j - x_j^*| < (d/2)^{\frac{1}{2}},\, 1 \le j \le n-1\},$$

$$m_{j,I,f}(t) = m_{j,I}(t) = (2d)^{\frac{1-n}{2}} \int_I \frac{\partial f}{\partial x_j}(x,t)\,dx,\ 1 \le j \le n-1,$$

$$m_{I,f}(t) = m_I(t) = (2d)^{\frac{1-n}{2}} \int_I f(x,t)\,dx\,.$$

Recall that $J_n = \{t : |t - t^*| < d/2\}$, and $Q_d(x^*, t^*) = I \times J_n$. We note for $1 \le j \le n-1$ and $t \in \bar{J}_n$, that

$$(5.9) \qquad |\, m_{j,I}(t) - m_{j,Q}\,| \le c a_2,$$

as follows from integration by parts and (1.4) . Again from (1.4) , we see that

$$(5.10) \qquad |m_Q - m_I(t)| \le c a_2\, d^{\frac{1}{2}},\ t \in \bar{J}_n\,.$$

Next from the usual BMO type argument (see [JN]) we deduce that

$$(5.11) \qquad |m_{j,Q} - m_{j,Q^*}| \le c \log(2 + \frac{r}{d})\, a_1\,.$$

Let $\eta_I(t) = \eta_{I,f}(t), \eta_Q = \eta_{Q,f}$, denote the points in R^{n-1} whose j th coordinates are $m_{j,I}(t)$, $m_{j,Q}$ respectively for $1 \le j \le n-1$ and $t \in \bar{J}_n$. Since f has distributional derivatives in the spatial variable, we observe from Morrey's lemma (see [S, ch 5]) that for almost every $(x,t) \in Q$, we have

(5.12)

$$|\, f(x,t) - m_I(t) - \langle\, \eta_I(t),\, x - x^* \,\rangle\,| \le c \int_I |\nabla_y f(y,t) - \eta_I(t)|\, |x-y|^{2-n}\, dy\,.$$

Using (5.9), (5.10), and integrating (5.12) over Q we obtain since $f \in BMO$ on 'rectangles' (see (1.17)) that

$$(5.13) \qquad \begin{aligned} &|Q|^{-1} \int_Q |\, f(x,t) - m_Q - \langle\, \eta_Q,\, x - x^* \,\rangle\,|\, dx dt \\ &\le c|Q|^{-1} \int_Q \int_I |\nabla_y f(y,t) - \eta_Q|\, |x-y|^{2-n}\, dx dy dt + c a_2\, d^{\frac{1}{2}} \\ &\le c\, d^{\frac{1}{2}}\, |Q|^{-1} \int_Q |\nabla_y f(y,t) - \eta_Q|\, dy dt + c a_2\, d^{\frac{1}{2}} \\ &\le c(a_1 + a_2)\, d^{\frac{1}{2}}\,. \end{aligned}$$

From (5.13), (5.11), and a well known argument on averages it follows that the function h defined by

$$(5.14) \qquad h(x,t) = f(x,t) - m_{Q^*} - \langle\, \eta_{Q^*},\, x - x_1 \,\rangle,\ (x,t) \in Q^*,$$

is uniformly Hölder continuous on Q^* in the space variables, for any exponent less than one. The same argument also yields

$$|h(x,t)| \leq c(a_1+a_2)\, r^{\frac{1}{2}}, \text{ when } (x,t)\in Q^*. \tag{5.15}$$

To prove (5.8) let $\tilde{Q} = Q_{2\rho}(x_1,t_1)$, and put

$$G(s) = \{(x,t)\in\tilde{Q} : |x-x_1|^2 + |\langle \eta_{\tilde{Q}}, x-x_1\rangle|^2 \leq s \text{ and } |t-t_1| \leq s\},$$

for $s>0$. Given $c_6>2$, let $r_1 = \frac{\rho}{c_6(1+a_1+a_2)^3}$, and $r_2 = c_6(1+a_1+a_2)^3\rho$. We claim that

$$G(r_1) \subset G \subset G(r_2), \tag{5.16}$$

for c_6 large enough. Indeed, to prove the lefthand inclusion in (5.16), we put $r=r_1$, and use (5.11), (5.15), and Hölder's inequality, to get for $(x,t)\in G(r_1)$, that

(5.17)

$$\begin{aligned} r_1 &\geq |x-x_1|^2 + |\langle \eta_{\tilde{Q}}, x-x_1\rangle|^2 \\ &\geq |x-x_1|^2 + \tfrac{1}{2}|\langle \eta_{Q^*}, x-x_1\rangle|^2 - c|\langle \eta_{\tilde{Q}} - \eta_{Q^*}, x-x_1\rangle|^2 \\ &\geq |x-x_1|^2 + \tfrac{1}{4}|f(x,t)-f(x_1,t_1)|^2 - c[h(x,t)-h(x_1,t_1)]^2 \\ &\qquad\qquad - c a_1^2 r_1 [\log(2+\tfrac{\rho}{r_1})]^2 \\ &\geq |x-x_1|^2 + \tfrac{1}{4}|f(x,t)-f(x_1,t_1)|^2 - c\,[\log(c_6(1+a_1+a_2))]^2 \frac{\rho}{c_6(1+a_1+a_2)}. \end{aligned}$$

Clearly (5.17), implies the lefthand inclusion in (5.16) for c_6 large enough. The righthand inclusion is proved similarly using the fact that $G\subset\tilde{Q}$. . We omit the details.

We now use (5.16), the triangle inequality, and (1.17) to prove the righthand inequality in (5.8). In fact,

$$\sigma(Q_\rho(X_1,t_1)) \leq \int_{G(r_2)} \sqrt{1+|\nabla_y f(y,s)|^2}\, dyds$$

$$\leq c\int_{\tilde{Q}} \sqrt{1+|\nabla_y f(y,s) - \eta_{\tilde{Q}}|^2}\, dydt + c(1+|\eta_{\tilde{Q}}|)|G(r_2)| \tag{5.18}$$

$$\leq c r_2^{\frac{n+1}{2}}.$$

To prove the lefthand inequality in (5.8), we consider two cases. First suppose that

$$|\eta_{\tilde{Q}}| \leq c_6^2(1+a_1+a_2)^4. \tag{5.19}$$

In this case we observe from (5.16) that

$$r_1^{\frac{n+1}{2}} \leq c c_6^2(1+a_1+a_2)^4 |G(r_1)| \leq c c_6^2(1+a_1+a_2)^4 \mu(Q_\rho(X_1,t_1)). \tag{5.20}$$

Hence the lefthand inequality in (5.8) is valid in this case, once we fix c_6.

If (5.19) is false, we need to use the inequality

$$\sup_{Q}[\,|Q|^{-1}\int_{Q}|\nabla_y f(y,s)\,-\,\eta_Q\,|^2\,dyds\,]\leq c\,a_1^2$$

whenever $Q = Q_d(x^*, t^*) \subset R^n$. This inequality follows from the well known fact that the square root of the left hand side defines an equivalent norm for BMO. Let $Q_1 = Q_{2r_1}(x_1, t_1)$. Then using this inequality, and Hölder's inequality, we obtain

(5.21)

$$\int_{G(r_1)}|\nabla_y f(y,s)\,-\,\eta_{Q_1}\,|\,dyds\;\leq\;\left[\int_{Q_1}|\nabla_y f(y,s)\,-\,\eta_{Q_1}\,|^2\,dyds\right]^{\frac{1}{2}}\;|G(r_1)|^{\frac{1}{2}}$$

$$\leq\;c a_1(\eta_{\tilde{Q}})^{-\frac{1}{2}}\;r_1^{\frac{n+1}{2}}\;.$$

Using (5.16), (5.21), and (5.11) , we conclude that if (5.19) is false, then

$$\sigma(Q_\rho(X_1,t_1))\geq\int_{G(r_1)}|\,\nabla_y f(y,s)\,|\,dyds$$

$$\geq\;(|\eta_{\tilde{Q}}|\,-\,|\eta_{\tilde{Q}}-\eta_{Q_1}|)|G(r_1)|\,-\int_{G(r_1)}|\nabla_y f(y,s)\,-\,\eta_{Q_1}\,|\,dyds \tag{5.22}$$

$$\geq\;(1\;-\;\frac{c\log[c_6(1+a_1+a_2)]}{c_6(1+a_1+a_2)}\,)\,r_1^{\frac{n+1}{2}}\;.$$

Hence the right hand inequality in (5.8) is valid in this case also. From (5.22), (5.20), and (5.18), we deduce that (5.8) is true.

In the rest of the proof of Theorem 3, we let A denote a positive constant depending only on $n, \delta, a_1, a_2,$ not necessarily the same at each occurence. Let $H = h$ in (5.14) when $r = 2\rho$ and extend H to R^n by letting H be equal to the righthand side of (5.14) in $R^n \setminus \bar{Q}_{2\rho}(x_1, t_1)$. Also, let $M(|\nabla_x H|)$ denote the maximal function of $\chi\,|\nabla_x H\,|$ on 'rectangles' defined by

$$M(|\nabla_x H|)(x^*,t^*)\;=\;\sup_{d>0}\;[\,|Q|^{-1}\int_{Q}\chi\,|\nabla_x H(x,t)|\,dxdt\,],$$

where χ is the characteristic function of $Q_{4\rho}(x_1, t_1)$ and $Q = \bar{Q}_d(x^*, t^*)$, is as above. Next we construct the measure ν in (4.19)(C) relative to σ, $Q_\rho(X_1, t_1)$. To do this, choose a sequence $\{Q^j\}$ of closed disjoint rectangles contained in $\zeta^{-1}(Q_\rho(X_1, t_1))$, with sidelength, $s = \frac{\rho^{\frac{1}{2}}}{A(1+|\eta_{\tilde{Q}}|)}$ in the spatial coordinates and s^2 in the t coordinate, disjoint interiors, and the property that

$$A\,|\cup Q^j\,|\geq(1\,+\,|\eta_{\tilde{Q}}|)^{-1}\,\rho^{\frac{n+1}{2}}\,.$$

Existence of this sequence follows easily from (5.16) . We also choose $E^j \subset Q^j$, for each j, so that (1.15), (1.16), hold with $\bar{Q}_d(x^*, t^*)$, E_1 replaced by Q^j, E^j

respectively , and $\alpha = \frac{1}{2}$. We put $E = \cup E^j$, and observe from (1.15), (1.16), and (1.4) that for $(x,t) \in E$,

$$\int_K \frac{(f(x,t)-f(x,s))^2}{|t-s|^2}\, ds \le A \log(2 + |\eta_{\tilde{Q}}|), \tag{5.23}$$

where $K = \{t : |t - t_1| < 4\rho\}$. Moreover,

$$\rho^{\frac{n+1}{2}} \le A(1 + |\eta_{\tilde{Q}}|)\, |E|. \tag{5.24}$$

We shall need the following lemma.

Lemma 5.1 *If* $A_1 = A_1(n, \delta, a_1, a_2)$, *is large enough, there exists a closed set* E_2 *and a compactly supported* $\phi : R^n \to R$ *satisfying (1.1) - (1.4), with* a_1, a_2 *replaced by* $A \log(2 + |\eta_{\tilde{Q}}|)$. *Also,*

(A) $E_2 \subset E \cap \{(x,t) : M(|\nabla_x H|)(x,t) \le A_1 \log(2 + |\eta_{\tilde{Q}}|)\}$,

(B) $A_1 |\zeta(E_2)| \ge \rho^{\frac{n+1}{2}}$,

(5.25) (C) $\phi = H$ *on* E_2,

(D) $$\int_K \frac{(\phi(x,t)-\phi(x,s))^2}{|t-s|^2}\, ds \le A_1 (1 + |\eta_{\tilde{Q}}|)\, [\log(2 + |\eta_{\tilde{Q}}|)]^2,$$
whenever $(x,t) \in E_2$.

Proof. To prove (5.25)(A), we shall use the John - Nirenberg type distribution inequality

$$|\{(x,t) : M(|\nabla H|)(x,t) > \lambda\}| \le A e^{-\frac{\lambda}{A}} |\tilde{Q}|.$$

A proof of this inequality for functions in BMO on 'cubes' can be found in [JN] . The proof is essentially unchanged for functions in BMO on 'rectangles' . If $\lambda = A_1 \log(2 + |\eta_{\tilde{Q}}|)$, and A_1 is large enough, we see from the above inequality and (5.24), that there exists E_1, satisfying (5.25)(A) with E_2 replaced by E_1, and

$$\rho^{\frac{n+1}{2}} \le A(1 + |\eta_{\tilde{Q}}|)\, |E_1|. \tag{5.26}$$

We note that E_2 will be chosen to be a certain subset of E_1 with $|E_2| = \frac{1}{2}|E_1|$.

To construct ϕ we observe that $E_1 \subset Q_\rho(x_1, t_1)$. Then as in section 2, we find that $\tilde{Q} \setminus E_1 = \cup Q_i$, where $\{Q_i\}_1^\infty$ are closed Whitney 'rectangles', with disjoint interiors, and the following properties: Q_k has its center at (y_k, t_k), sidelength r_k in the x_j direction, $1 \le j \le n-1$, and r'_k in the t direction, where

$$c^{-1} (r'_k)^{\frac{1}{2}} \le r_k \le c\, (r'_k)^{\frac{1}{2}},\ k = 1, 2, \ldots \tag{5.27}$$

for some $c \ge 2$. Also, if d denotes the nonisotropic distance function with $\alpha = \frac{1}{2}$, so that

$$d(Q_k, E_1) = \min\{|x - y| + |s - t|^{\frac{1}{2}} : (x,t) \in Q_k,\ (y,s) \in E_1\},$$

then

$$(5.28)\qquad c\, d(Q_k, E_1) < \text{ diam } Q_k < \frac{1}{100\sqrt{n-1}}\, d(Q_k, E_1),\ k = 1, 2, \ldots .$$

Let ψ_1, ψ_2, be as in section 2 . Then, as in section 2, we put

$$v(x,t) = \sum_k \psi_1(\tfrac{t-t_k}{r'_k})\, \psi_2(\tfrac{x-y_k}{r_k\sqrt{n-1}}),$$

$$v_i(x,t) = v(x,t)^{-1}\, \psi_1(\tfrac{t-t_i}{r'_i})\, \psi_2(\tfrac{x-y_i}{r_i\sqrt{n-1}})$$

and note that $\{v_i\}$ is a C^∞ partition of unity for $\tilde{Q}\setminus E_1$. From (5.27) and (5.28) we deduce, for fixed k and $T = \{i:\ \text{supp } v_i \cap \bar{Q}_k\} \neq \emptyset$, that

$$(5.29)\qquad \text{card } T \le c\,,$$

$$(5.30)\qquad c^{-1} r_l \le r_k \le c\, r_l,\ \ l \in T.$$

Put $\phi(x,t) = \sum_k H(y_k, t_k)\, v_k$ when $(x,t) \in \tilde{Q}\setminus E_1$ and $\phi = H$ on E_1 . For k, T, as above we claim that

$$(5.31)\qquad |\,\phi(x,t) - H(x,t)\,| \le A\ \log(2 + |\eta_{\tilde{Q}}|)(r'_k)^{\frac{1}{2}},\ (x,t) \in Q_l\,,$$

when $l \in T$. To prove this claim, observe from (5.27) and (5.30) that if c is large enough and $Q' = Q_{cr'_k}(x_k, t_k)$, then $Q_l \subset Q'$ whenever $l \in T$. Since H also satisfies (1.17) and (1.4) on subcubes of $Q_{4\rho}(x_1, t_1)$, we can repeat the argument leading to (5.15), to get

$$(5.32)\qquad |\,H(x,t) - m_{Q',H} - \langle \eta_{Q',H}\,,\, x - x_k\rangle\,| \le A(r'_k)^{\frac{1}{2}}$$

when $(x,t) \in Q'$. Also in view of (5.25)(A), we have

$$(5.33)\qquad |\,\eta_{Q',H}\,| \le A\log(2 + |\eta_{\tilde{Q}}|)\,.$$

From (5.32) and (5.33) we conclude that

$$|\phi(x,t) - H(x,t)| \le \sum_{l\in T} |\,H(x_l, t_l) - H(x,t)\,| \le A\log(2 + |\eta_{\tilde{Q}}|)(r'_k)^{\frac{1}{2}}\,.$$

Hence (5.31) is true.

Similarly, we deduce from (5.31) - (5.33), as in section 2, that ϕ is uniformly Lipschitz and Hölder $\frac{1}{2}$ on $\tilde{Q}$ in the sense of (1.1) and (1.4) with norms at most $A\ \log(2 + |\eta_{\tilde{Q}}|)$. Also

$$(5.34)\qquad |\,\frac{\partial\phi}{\partial t}\,(x,t)\,| \le A\log(2 + |\eta_{\tilde{Q}}|)(r'_k)^{-\frac{1}{2}}\,, (x,t) \in Q_k\,.$$

We extend the definition of ϕ to R^n as in section 2 (see the discussion following (2.32)) by requiring that ϕ be constant on certain line segments of $R^n \setminus \tilde{Q}$. We also denote this extension by ϕ. We now repeat the argument following (2.19) with ϕ playing the role of $\tilde{\phi}$ and H playing the role of ϕ_2. The proof is essentially unchanged, even though H is not necessarily Lipschitz in the space variables, thanks to (5.25)(A) and (5.31) - (5.34). We obtain that ϕ satisfies (1.2) and (1.3). Finally, we multiply ϕ by an appropriate cutoff function as in section 2, in such a way that that this function agrees with ϕ on $\tilde{Q}$ and properties (1.1)-(1.4) are preserved for the resulting function. We also denote the new function by ϕ. Finally we choose E_2 to be any subset of E_1 such that (5.25)(D) holds and $|E_2| = \frac{1}{2}|E_1|$. Existence of E_2 for A_1 large enough follows from the Carleson measure condition (1.14) on $\tilde{Q}$, (5.26), and weak type estimates. (5.25) (B) can be proved using (5.26), (5.16), and an argument similar to the argument used in proving the left hand inequality in (5.8). The proof of Lemma 5.1 is now complete. □

Next we let

$$\phi_1(x,t) = \phi(x,t) + m_{\tilde{Q},f} + \langle \eta_{\tilde{Q},f}\, , x - x_1 \rangle,\ (x,t) \in R^n ,$$

and put $\zeta_1(x,t) = (x, \phi_1(x,t), t)$, when $(x,t) \in R^n$, where ϕ is as in Lemma 5.1. Also, set $\mathcal{S}_1 = \{\zeta_1(x,t) : (x,t) \in R^n\}$. To define ν we consider two cases. First if (5.19) is true, with $c_6^2(1+a_1+a_2)^4$ replaced by A_2, we let

$$\nu(F) = \int_{\zeta_1^{-1}(\mathcal{S}_1 \cap F)} \sqrt{1 + |\nabla_y \phi_1(y,s)|^2}\, dyds,$$

when F is a Borel subset of R^{n+1} . From (5.25)(B), the fact that $\phi_1 = f$ on E_2, and Theorem 1, we deduce that ν satisfies the hypotheses of Lemma 4.2 with E replaced by $\zeta_1(E_2), \alpha = \frac{1}{2}, l = n,$ and $k = n-1$.

On the other hand if (5.19) is false, we choose a rotation U of R^{n-1} in such a way that if $x = Ux'$, then $\eta_{\tilde{Q}} = |\eta_{\tilde{Q}}|\, Ue_1$, where $e_1 = (1,...,0)$ in the rotated coordinate system . Put $\phi_2(x') = \phi_1(Ux')$ and note from Lemma 2 that

$$\text{(5.35)}\quad \begin{aligned} &(i) && \left|\frac{\partial\phi_2}{\partial x_1'} - |\eta_{\tilde{Q}}|\right| \le A_1 \log(2 + |\eta_{\tilde{Q}}|), \\ &(ii) && \left|\frac{\partial\phi_2}{\partial x_j'}\right| \le A_1 \log(2 + |\eta_{\tilde{Q}}|),\ 2 \le j \le n-1. \end{aligned}$$

We now proceed as in section 3. Let $\theta = -\arctan(|\eta_{\tilde{Q}}|)$, $-\frac{\pi}{2} < \theta < 0$, and set

$$\text{(5.36)}\quad \begin{aligned} x_1' &= \cos\theta\, \bar{x}_1 + \sin\theta\, \phi_3(\bar{x}_1, ..., \bar{x}_{n-1}, t), \\ x_j' &= \bar{x}_j, 2 \le j \le n-1, \\ \phi_2(x_1', ..., x_{n-1}', t) &= -\sin\theta\, \bar{x}_1 + \cos\theta\, \phi_3(\bar{x}_1, ..., \bar{x}_{n-1}, t). \end{aligned}$$

Using (5.35) and arguing as in (3.17) we see that if A_2 is large enough in (5.19), then ϕ_3 exists satisfying (5.36) and

$$\text{(5.37)}\quad |\nabla_{\bar{x}}\, \phi_3|(\bar{x},t) \le \frac{A \log(2 + |\eta_{\tilde{Q}}|)}{1 + |\eta_{\tilde{Q}}|},$$

whenever $(\bar{x},t) \in R^n$. Using (5.37), Lemma 5.1, and arguing as in (3.23), we also get

$$\text{(5.38)}\quad |\phi_3(\bar{x},t) - \phi_3(\bar{x},s)| \le \frac{A[\log(1 + |\eta_{\tilde{Q}}|)]^2}{1 + |\eta_{\tilde{Q}}|} |s-t|^{\frac{1}{2}},\ (\bar{x},t), (\bar{x},s) \in R^n .$$

Moreover if $\tilde{E}$ denotes the image of E_2 under the transformation $(x,t) \to (x',t) \to (\bar{x},t)$, then from our choice of θ and (5.25)(D), we deduce as in (3.23), (3.25), that

$$\int_K \frac{(\phi_3(\bar{x},t)-\phi_3(\bar{x},s))^2}{|s-t|^2}\,ds \le A\frac{[\log(2+|\eta_Q|)]^2}{1+|\eta_Q|^2}\int_K \frac{(\phi(x,t)-\phi(x,s))^2}{|s-t|^2}\,ds \tag{5.39}$$

$$\le A\,\frac{[\log(2+|\eta_Q|)]^4}{1+|\eta_Q|}$$

when $(\bar{x},t) \in \tilde{E}$. From (5.37) - (5.39) we see that once again Lemma 2.1 can be applied with ϕ_3 playing the role of both ϕ_1, ϕ_2, and $F = \tilde{E}$. Doing this, we get ϕ_4 satisfying (1.1) - (1.4) with a_1, a_2, replaced by $\frac{A\,[\log(2+|\eta_Q|)]^2}{1+|\eta_Q|^{\frac{1}{2}}}$ and $\phi_3 = \phi_4$ on $\tilde{E}$.

Given $(\bar{x},t) \in R^n$, let $\hat{x} = (\hat{x}_1, ..., \hat{x}_{n-1})$, where $\hat{x}_1(\bar{x},t) = \cos\theta\,\bar{x}_1 + \sin\theta\,\phi_4(\bar{x},t)$, and $\hat{x}_j = \bar{x}_j$, for $2 \le j \le n-1$. Define $\zeta_2 : R^n \to R^{n+1}$ by

$$\zeta_2(\bar{x},t) = (U\hat{x}(\bar{x},t),\ -\sin\theta\,\bar{x}_1 + \cos\theta\,\phi_4(\bar{x},t),\ t).$$

Set $\mathcal{S}_2 = \{\zeta_2(\bar{x},t)) : (\bar{x},t) \in R^n\}$, and let

$$\nu(F) = \int_{\zeta_2^{-1}(F\cap\mathcal{S}_2)} \sqrt{1 + |\nabla_{\bar{y}}\,\phi_4(\bar{y},s)\,|^2}\, d\bar{y}ds.$$

We note from (5.36), as in the display preceding (3.21), that

$$\sqrt{1+|\nabla_x\phi_1|^2}dxdt = \sqrt{1+|\nabla_{\bar{x}}\phi_3|^2}d\bar{x}dt\,. \tag{5.40}$$

From (5.40) and our construction it follows that

$$L = \zeta(E_2) = \zeta_2(\tilde{E}) \subset \mathcal{S} \cap \mathcal{S}_2$$

and $\nu|_L = \sigma|_L$. From this fact and (5.25)(B), we find that

$$A\,\nu(L) \ge \rho^{\frac{n+1}{2}}\,. \tag{5.41}$$

Using (1.1) - (1.4) for ϕ_4 and Theorem 1, it is also easily deduced that the corresponding operators in Lemma 4.2 , defined relative to ν, are bounded by A. From this deduction and (5.41), we conclude that ν also satisfies the hypotheses of Lemma 4.2 when (5.19) is false for sufficiently large A_2. Since we have considered the case when (5.19) is true, and $(X_1,t_1) \in S$, $\rho > 0$, are arbitrary, we conclude from Lemma 4.2 that Theorem 3 is true. □

6. Further Results.

In this section we introduce parabolic analogues of David ω regular surfaces and state a theorem concerning boundedness of the corresponding parabolic singular integrals on our surfaces. In order to state our theorem we shall need some more definitions and notation. Let $1 \le k \le n-1$ be a positive integer and let $\zeta = (\zeta_1, \zeta_2, ..., \zeta_n, t)$ be a function from R^{k+1} into R^{n+1} . Let $\bar{Q}_d(x^*,t^*)$ be as in (1.11) with $k+1 = n$ and suppose that for each $(x^*,t^*) \in R^{k+1}$, $d > 0$, and some δ, $0 < \delta < 1$, there exists $E_1 \subset \bar{Q}_d(x^*,t^*)$ such that (1.15), (1.16) hold with

$n = k+1$ and $f = \zeta_j$, $1 \le j \le n$. We also assume that each component of ζ has distributional partial derivatives and for almost every $(x,t) \in R^{k+1}$,

$$(6.1) \qquad |\nabla_x \zeta_j(x,t)| \le \omega(x,t)^{\frac{1}{k}}, 1 \le j \le n.$$

In (6.1), ω is a positive Lebesgue measurable, locally integrable function defined on R^{k+1}. Put

$$\tilde{\omega}(F) = \int_F \omega(x,t)dxdt,$$

whenever $F \subset R^{k+1}$ is Lebesgue measurable. We say that $\tilde{\omega}$ is a doubling measure if there exists $c > 0$ such that whenever $0 < d < \infty$ and $(x^*,t^*) \in R^{k+1}$,

$$(6.2) \qquad \tilde{\omega}(Q_{2d}(x^*,t^*)) \le c\,\tilde{\omega}(\bar{Q}_d(x^*,t^*)).$$

We say that $\tilde{\omega} \in A_\infty(R^{k+1})$ on 'rectangles' provided there exists positive constants ϵ, c such that whenever $F \subset \bar{Q}_d(x^*,t^*)$ is Borel, we have

$$(6.3) \qquad \frac{\tilde{\omega}(F)}{\tilde{\omega}(\bar{Q}_d(x^*,t^*))} \le c\left[\frac{|F|}{|\bar{Q}_d(x^*,t^*)|}\right]^\epsilon,$$

for all $d > 0$ and $(x^*,t^*) \in R^{k+1}$. Let $\mathcal{S} = \{\zeta(x,t) : (x,t) \in R^{k+1}\}$ and let μ be the induced Borel measure with supp $\mu \subset \mathcal{S}$, defined by

$$\mu(F) = \tilde{\omega}\,[\zeta^{-1}(F \cap \mathcal{S})],$$

when $F \subset R^{n+1}$, is a Borel set. We suppose for each $d > 0$ and $(X^*,t^*) \in \mathcal{S}$, that

$$(6.4) \qquad \mu(Q_d(X^*,t^*)) \le c\,d^{\frac{k}{2}+1}.$$

We say that $\mathcal{S}$ is a parabolic $(\omega, k+1)$ regular surface if (1.4), (1.15) - (1.16), and (6.1) - (6.4) hold for ω, ζ as described above.

Define singular integral operators on S as in section 1, by

$$P_{j,\epsilon}\, g(X,t) = \int W_{j,\epsilon}(X-Y, t-s)\, g(Y,s)\, d\mu(Y,s)$$

and

$$\bar{P}_j\, g(X,t) = \sup_{\epsilon>0} |P_{j,\epsilon}\, g(X,t)|,$$

when $(X,t) \in R^{n+1}$ and $g \in C_0^\infty(R^{n+1})$. The following theorem is a parabolic version of the main theorem in [D1].

Theorem 4 *Let $\mathcal{S}$ be a parabolic $(\omega, k+1)$ regular surface. Then for $1 \le j \le 2n$ the operator $P_{j,\epsilon}$ can be extended to a bounded operator on $L^p(\mu)$ with norm constants $\hat{c}_j$, depending only on n, p, and the constants in (1.4), (1.14)-(1.15), (6.1)-(6.4). Moreover, (1.8), (1.9), of Theorem 1 are valid with (x,t) replaced by $(X,t) \in R^{n+1}$; K_j by P_j, c^j by $\hat{c}_j$, and $\|\cdot\|_p$ by $\|\cdot\|_{p,\mu}$, whenever $h \in L^p(\mu)$.*

Proof. We sketch the proof of Theorem 4, when $\omega = c$. In this case we see from (6.1) that ζ_j, $1 \le j \le k$, is uniformly Lipschitz in the space variable in the sense of (1.1). If also $k = 1$, $n = 2$, then a proof of this theorem can be given, using Lemma 2.1, the rising sun argument (as in section 3), Lemma 4.2, and Theorem 2 (compare with [D3, section 4]). Indeed, in this case given $\rho > 0$ and $(x_1, t_1) \in R^2$, it follows from (6.1), (6.4), Lemma 2.1, and the rising sun argument that there exists a function $h: R^2 \to R$ which is bilipschitz in x, uniformly in t, and satisfies (1.2)-(1.4) with constants depending only on the constants for ζ in (1.4),(1.15) - (1.16), and (6.1) - (6.4). Moreover h is equal to some component (say ζ_1) of ζ on a set $E \subset Q_\rho(x_1, t_1)$, with $A|E| \ge |Q_\rho(x_1, t_1)|$, where A depends only on the data. For fixed $t \in R$ let $p(\cdot, t)$ denote the inverse of $h(\cdot, t)$. Given $(x', t) \in R$, we consider $q(x', t) = \zeta_2(p(x', t), t)$. Using (1.1) for p, and (1.4), (1.13), (1.14), for h, it is easily shown that (1.15) holds with $f = q$, whenever $d > 0$ and $(x^*, t^*) \in R^{k+1}$, for some $E_1 \subset \bar{Q}_d(x^*, t^*)$ satisfying (1.16) with appropriate constants. Also, if

$$E' = \{(x', t) : (p(x', t), t) \in E\},$$

then $A|E'| \ge |Q_\rho(x_1, t_1)|$. Let $\zeta'(x', t) = (x', q(x', t), t)$ and put $\mathcal{S}' = \{\zeta'(x', t) : (x', t') \in R^2\}$. Defining ν as in section 4 relative to ζ', using the above facts, and Theorem 2, we see that ν satisfies the hypotheses of Lemma 4.2 with E replaced by $\zeta(E)$. Applying Lemma 4.2, it follows that the above Theorem is true in case $\omega = c$, $n = 2$, $k = 1$.

To handle the case where $\omega = c$, $n > 2$, $1 \le k \le n - 1$, suppose $\zeta_j = x_j$ for l distinct integers j, $1 \le j \le k$. Suppose, for example, that $\zeta_j = x_j, 1 \le j \le l$. If $k = l$, we need to show that Theorem 1 remains valid when $|f(x,t) - f(y,s)|^2$ is replaced by

$$\sum_{i=k+1}^{n} |\zeta_i(x,t) - \zeta_i(y,s)|^2$$

in the exponential term of the kernels of the operators in this theorem. This can be done using the arguments in this paper, essentially unchanged, once we generalize Theorem A to operators with kernels modified as above. Now, it is not difficult to modify the induction arguments in [LM1] to include these operators. In fact we originally proved such a theorem. However, to clarify somewhat the complicated induction arguments in the proof, we opted for the more select group of operators which appear in Theorem A. To handle the case when $l < k$, we use a simplified version of the ingenious argument in [D1, section 4]. Assume by way of induction that we have proved Theorem 4 whenever, $l \ge m + 1$, $0 \le m \le k - 1$, for all $n \ge 2$. If $l = m$ we use Lemma 2.1, the rising sun argument, (6.1), and (6.4), as in the case $n = 2$, $k = 1$, to deduce for given $\rho > 0$ and $(x_1, t_1) \in R^{k+1}$, that there exists a function $h: R^{k+1} \to R$ with the following property: for fixed $x_j, j \ne m+1$, and t, the function $x_{m+1} \to h(x,t)$ is bilipschitz as a function of x_{m+1} uniformly in t, and for varying x, t satisfies (1.1)-(1.4) with constants depending only on the constants for ζ in (1.15) - (1.16) and (6.1) - (6.4). Moreover h is again equal to some component (say ζ_{m+1}) of ζ on a set $E \subset Q_\rho(x_1, t_1)$, with $A|E| \ge |Q_\rho(x_1, t_1)|$, where A depends only on the data. For fixed x_j, $j \ne m + 1$, and t, let p be the inverse of the above function. Then p is defined on R^{k+1} and either increasing or decreasing on lines parallel to the x_{m+1} axis. Given $(x', t) \in R^{k+1}$ define $\beta: R^{k+1} \to R^k$, by $\beta_j(x', t) = x'_j$, $j \ne m + 1$, $1 \le j \le k$, and $\beta_{m+1}(x',t) = p(x',t)$. Again we consider $q_j(x',t) = \zeta_j(\beta(x',t),t)$, $m+2 \le j \le k$, $(x',t) \in R^{k+1}$. As in the case $n = 2, k = 1$, we find that q_j satisfies (1.4),(1.15), and (1.16) whenever $(x^*,t^*) \in R^{k+1}$, $d > 0$. Let $\zeta' : R^{k+1} \to R^{n+2}$ be defined as follows: put $\zeta'_j(x',t) = x'_j$, $1 \le j \le m+1$, and put $\zeta'_j(x',t) = q_j(x',t)$ for $m+2 \le j \le n$. Let $\zeta'_{n+1}(x',t) = \zeta_{m+1}(\beta(x',t),t) - x'_{m+1}$, and set $\zeta'_{n+2}(x',t) = t$. Set $\mathcal{S}' = \{\zeta'(x',t) : (x',t) \in R^{k+1}\}$ and define ν relative to $\zeta', \mathcal{S}'$,as previously. Then it is easily shown that ν satisfies (6.4) and hence $\mathcal{S}' \subset R^{n+2}$ is a parabolic $(c, k+1)$ regular surface for some c. Since ζ' satisfies the induction hypothesis, we see that Theorem 4 is valid for the singular integral operators defined on $\mathcal{S}'$. Next we identify $\mathcal{S}$ with $\{(x,0,t) : (x,t) \in R^{n+1}\}$, which is a subset of R^{n+2}. Then ν and μ agree on $\zeta(E)$ and so again it follows from Lemma 4.2 that Theorem 4 holds for the singular integral operators on $\mathcal{S}$. Hence, Theorem 4 is valid whenever $l \ge m$. By induction we now get Theorem 4 in case $\omega = c$. The proof of Theorem 4 for an arbitrary ω satisfying (6.2), (6.3) follows from the case $\omega = c$ as in [D1]. □

We plan to give more details of the proof of Theorem 4 in a future paper, where we also hope to obtain generalizations of the main theorems in [DS1], [DS2].

We note that we have been unable to use the technique in [D2] (see also [Js]) to prove Theorem 4, primarily because in order to insure that (1.15) is true for functions similar to ζ', we appear to need a global inverse. Finally, we remark that a purely geometric characterization of surfaces, which have big pieces of graphs of functions f satisfying (1.1)-(1.4), as in [DJ], appears difficult because of the $\frac{1}{2}$ derivative condition in t.

7. References

[D1] G. David, *Opérateurs d'intégrale singulière sur les surfaces régulières* , Ann. Sci. Ec. Norm. Sup. (4) **21**(1988), 225 - 258 .

[D2] G. David, *Morceaux de graphes lipschitziennes et intégrales singulières sur un surface,* Rev. Mat. Iberoamericana **4**(1988), 73 - 114.

[D3] G. David, *Wavelets and singular integrals on curves,* Lecture Notes in Math. **1465**, Springer-Verlag, 1991 .

[DJ] G. David and D. Jerison, *Lipschitz approximations to hypersurfaces, harmonic measure, and singular integrals,* Indiana Math. J **39** (1990), 831 - 845 .

[DS1] G. David and S. Semmes, *Singular integrals and rectifiable sets in* R^n *; au-delà des graphes lipschitziens,* Astérisque **193**, 1991 .

[DS2] G. David and S. Semmes, *Quantitative Rectifiability and Lipschitz mappings,* to appear.

[JN] F. John and L. Nirenberg, *On functions of bounded mean oscillation,* Comm. Pure and Applied Math. **14** (1961), 415 - 426 .

[Js] P. Jones, *Lipschitz and bilipschitz functions,* Revista Matematica Iberoamericana **4**(1988), 115-122 .

[J] J-L Journé, *Calderón - Zygmund operators, pseudodifferential operators and the Cauchy integral of Calderón,* Lecture Notes in Math. **994** , Springer-Verlag 1983.

[LM] J.L. Lewis and M.A.M. Murray, *Absolute Continuity of Parabolic Measure,* in Partial Differential Equations with Minimal Smoothness and Applications, IMA Volumes In Mathematics And Its Applications **42** (1992), Springer-Verlag, 173-189.

[S] E.M. Stein, *Singular Integrals and Differentiability Properties of Functions,* Princeton Univ. Press, Princeton, 1970.

[St] R.S. Strichartz, *Bounded mean oscillation and Sobolev spaces,* Indiana U. Math. J. **29** (1980), 539-558.

[Z] A. Zygmund, *Trigometric series,* Cambridge University Press, 1968.

CHAPTER III

Absolute Continuity and Dirichlet- Neumann Problems

1. Introduction

Recall that in chapters 1 and 2 we considered a function $f: R^n \to R$ with the following properties:

$$(1.1) \qquad |f(x,t) - f(y,t)| \le a_1 |x-y|, \; x, y \in R^{n-1}, \; t \in R,$$

$$(1.2) \qquad f(x,t) = I_{\frac{1}{2}}(b(x,.))(t) = \int_R |s-t|^{-\frac{1}{2}} b(x,s)ds,$$

where $x \in R^{n-1}$ is fixed and $b(x,.)$ is of bounded mean oscillation on R with BMO norm satisfying

$$(1.3) \qquad \| b(x,.) \|_* \le a_2 .$$

From (1.2) - (1.3) it follows easily that

$$(1.4) \qquad |f(x,s) - f(x,t)| \le c a_2 |s-t|^{\frac{1}{2}}, \; s, t \in R .$$

Here c as elsewhere in this chapter denotes an absolute constant depending only on n, not necessarily the same at each occurence. More generally, $c(b_1, b_2, \dots, b_m)$ will denote a constant depending only on $b_1, b_2, \dots, b_m$. If $Z = (z, z_n)$, $z \in R^{n-1}$, $z_n \in R$, we again let $D \subset R^{n+1}$ be the domain

$$D = \{(Z,t) : z_n > f(z,t)\}.$$

If $F \subset \partial D$ is a Borel set, let $\tilde{\omega}(Z,t,F,D)$ be parabolic or caloric measure of F evaluated at (Z,t). That is, $\tilde{\omega}$ is the bounded solution to the Dirichlet problem for the heat equation in D with boundary values 1 on F and 0 on $\partial D \setminus F$ in the sense of Perron-Wiener-Brelot. Next we identify sets on ∂D with sets in R^n by way of the mapping $\rho: R^n \to \partial D$ defined by

$$\rho(z,t) = (z, f(z,t), t), \; (z,t) \in R^n$$

and define ω a probability measure on R^n corresponding to $\tilde{\omega}$ by

$$\omega(Z,t,E) = \tilde{\omega}(Z,t,\rho(E),D), \; E \text{ Borel} \subset R^n .$$

As in chapter 2 we let $Q_d(x^*,t^*)$ be the rectangle

$$(1.5) \qquad Q_d(x^*,t^*) = J_1 \times J_2 \times \dots J_n$$

where $J_i = \{x \in R^n : |x_i - x_i^*| < (d/2)^{\frac{1}{2}}\}$ for $1 \le i \le n-1$ and $J_n = \{t : |t-t^*| < d/2\}$. We shall often write Q^* for $Q_d(x^*,t^*)$. Most of the effort in sections 3 - 5 is devoted to proving the following theorem.

Theorem 1. *Let f satisfy (1.1)-(1.3) and D, Q^* be as above. Let $E \subset Q^*$ be a Borel set and put $X^* = (x^*, f(x^*,t^*))$. Let $A \ge 2$, and suppose $(X,t) \in D$ with*

$t - t^* \geq d$ while $A(t - t^*) \geq |X - X^*|^2$. Then there exists $c_1 = c_1(a_1, a_2, A, n) \geq 2$, and $\beta = \beta(a_1, a_2, n)$, $0 < \beta < \frac{1}{2}$ such that

$$c_1^{-1}\left[\frac{\omega(X,t,E)}{\omega(X,t,Q^*)}\right]^{\frac{1}{\beta}} \leq \frac{|E|}{|Q^*|} \leq c_1\left[\frac{\omega(X,t,E)}{\omega(X,t,Q^*)}\right]^{\beta}. \tag{1.6}$$

We note for fixed (X,t) that (1.6) implies $\omega(X,t,\cdot)$ satisfies an A_∞ type condition(see [T, ch 9] for definitions) with respect to Lebesgue measure on certain ‘ rectangles ’ of the form Q^*.

Using Theorem 1, a comparison lemma of Brown (see section 4 for a statement of this lemma) and our work on parabolic versions of the David buildup scheme in chapter 2, we shall show that the hypotheses in Theorem 1 can be considerably relaxed to get a stronger theorem. Indeed suppose $f: R^n \to R$ has distributional partial derivatives, $\frac{\partial f}{\partial x_i}$, $1 \leq i \leq n-1$, which satisfy

$$\frac{\partial f}{\partial x_i} \in BMO(\ R^n\ ),\ 1 \leq i \leq n-1, \tag{1.7a}$$

$$f(x,t) = I_{\frac{1}{2}} * b(x,\cdot)(t),\ b \in BMO(\ R^n\ ). \tag{1.7b}$$

In (1.7), BMO(R^n) is the space of functions of bounded mean oscillation on ‘ rectangles ’ in R^n of the form (1.6) defined as follows: $k \in BMO(\ R^n\ )$ if for $d > 0$, $(x^*, t^*) \in R^n$, and

$$m_{Q^*} = m_{Q^*}(k) = |Q^*|^{-1}\int_{Q^*} k(x,t)\,dxdt\,,$$

we have

$$\int_{Q^*} |\,k(x,t) - m_{Q^*}\,|\,dxdt \leq \gamma\,|Q^*|,$$

where $|Q^*|$ denotes the Lebesgue n measure of Q^* and $0 < \gamma < \infty$. The infimum of all such γ for which the above inequality holds for all ‘ rectangles ’ in R^n is called the BMO norm of k.

Define D relative to f as above, let $\nabla_x f = (\frac{\partial f}{\partial x_1}, \ldots, \frac{\partial f}{\partial x_{n-1}})$ and set

$$\sigma(F) = \iint_{\rho^{-1}(F)} \sqrt{1 + |\nabla_x f|^2}\,dxdt$$

when $F \subset \partial D$. For $(X^*, t^*) = (x^*, f(x^*, t^*), t^*)$ as in Theorem 1, we let $Q_d(X^*, t^*)$ be the $n+1$ dimensional rectangle defined as in (1.5) with $x^*, n-1$, replaced by X^*, n. If $\Delta = \Delta(X^*, t^*) = Q_d(X^*, t^*) \cap \partial D$, we prove

Theorem 2. *Let f satisfy (1.4), (1.7), and suppose that $E \subset \Delta$ is a Borel set. If A, (X,t), are as in Theorem 1, then there exists c_2, γ, such that*

$$c_2^{-1}\left[\frac{\tilde{\omega}(X,t,E)}{\tilde{\omega}(X,t,\Delta)}\right]^{\frac{1}{\gamma}} \leq \frac{\sigma(E)}{\sigma(\Delta)} \leq c_2\left[\frac{\tilde{\omega}(X,t,E)}{\tilde{\omega}(X,t,\Delta)}\right]^{\gamma}. \tag{1.8}$$

Here γ, $0 < \gamma < \frac{1}{2}$, depends only on n, a_2 in (1.4), and the BMO norms in (1.7) while c_2 depends on these quantities and also on A.

Theorems 1 and 2 give A_∞ results on rectangles for parabolic measure. In section 6 we show that the hypotheses in Theorem 2 can once again be considerably weakened and we can still obtain absolute continuity results. We prove

Theorem 3. *Let f satisfy (1.4),(1.7a), and suppose*

$$\int_{J_n}\left(\int_{Q^*} \frac{(f(x,t)-f(x,s))^2}{(t-s)^2}\, dxdt\right) ds < \infty. \tag{1.9}$$

If (X,t) is as in Theorem 1, then $\sigma \circ \rho$ restricted to Q^* is absolutely continuous with respect to $\omega(X,t,\cdot)$.

Theorem 4. *Let f satisfy (1.4),(1.7a), and suppose (X,t) is as in Theorem 1. If E is a Borel subset of Q^*, $|E| = 0$, and $\omega(X,t,E) > 0$, then E is contained ω almost everywhere in the union of the sets :*

$$\{(x,t) \in Q^* : \textstyle\int_{J_n}\left(\int_{Q^*} \dfrac{(f(x,t)-f(x,s))^2}{(t-s)^2}\, dxdt\right) ds = \infty\}$$

$$\{(x,t) \in Q^* : M(|\nabla_x f|)(x,t) = \infty\}.$$

In Theorem 4, $M(|\nabla_x f|)$ denotes the Hardy Littlewood maximal function of $|\nabla_x f|$ on rectangles (see Lemma 6.7 for a definition). We note that Theorems 1-4 are higher dimensional analogues of the theorems in [LM]. To prove Theorems 1-4 we first use the method of layer potentials to study the L^p Dirichlet and Neumann problems for the heat equation in D when f satisfies (1.1)-(1.3) and a_1, a_2 are small. To this end recall from chapter 2 that the fundamental solution to the heat equation in $R^{n+1} \setminus \{0\}$ is given by

$$W(Z,\tau) = (4\pi\tau)^{-n/2} \exp\left(-\frac{|Z|^2}{4\tau}\right) \chi_{[0,\infty)}(\tau),\ Z \in R^n,\ \tau \in R,\ (Z,\tau) \neq (0,0).$$

Given $\epsilon > 0$, let $W_\epsilon(Z,\tau) = W(Z,\tau)$, when both $|Z| > \epsilon$ and $|\tau| > \epsilon^2$. Otherwise, put $W_\epsilon(Z,\tau) = 0$. Using the mapping $\rho : R^n \to \partial D$ we identify functions on ∂D with functions on R^n. Also as in chapter 2 we put $D_t = D \cap (R^n \times \{t\})$ for $t \in R$ and regard D_t as a domain in R^n, with outer unit normal

$$\nu_t(x) = \frac{(\nabla_x f(x,t), -1)}{\sqrt{|\nabla_x f(x,t)|^2 + 1}}$$

at $\rho(x,t) = (x, f(x,t), t)$. We define operators $\mathcal{L}, \mathcal{K}$ acting on a function $g : \partial D \to R$ by

$$(1.10)\quad \begin{aligned}\mathcal{L}_\epsilon g(x,t) &= \tfrac{1}{2}\int_{-\infty}^{t}\int_{\mathbb{R}^{n-1}} \tfrac{\langle \nu_s(y),(x-y,f(x,t)-f(y,s))\rangle}{(t-s)} \\ &\quad \cdot W_\epsilon(\rho(x,t)-\rho(y,s),t-s)\sqrt{1+|\nabla_y f(y,s)|^2}\,g(y,s)dyds, \\ \mathcal{K}_\epsilon g(x,t) &= \tfrac{1}{2}\int_{-\infty}^{t}\int_{\mathbb{R}^{n-1}} \tfrac{\langle \nu_t(x),(x-y,f(x,t)-f(y,s))\rangle}{(t-s)} \\ &\quad \cdot W_\epsilon(\rho(x,t)-\rho(y,s),t-s)\sqrt{1+|\nabla_y f(y,s)|^2}\,g(y,s)dyds,\end{aligned}$$

and

$$\mathcal{L}g(x,t) = \lim_{\epsilon\to 0}\mathcal{L}_\epsilon g(x,t), \quad \mathcal{K}g(x,t) = \lim_{\epsilon\to 0}\mathcal{K}_\epsilon g(x,t),$$

provided these limits exist. $\mathcal{L}$ is called the boundary double layer potential of g while $\mathcal{K}$ is the formal adjoint of $\mathcal{L}$. Next we define the single and double layer potentials of g on D, denoted S, L, respectively, by

$$(1.11)\quad \begin{aligned}Sg(X,t) &= \int_{-\infty}^{t}\int_{\mathbb{R}^{n-1}} W(X-\rho(y,s),t-s)\sqrt{1+|\nabla_y f(y,s)|^2}\,g(y,s)dyds, \\ Lg(X,t) &= -\tfrac{1}{2}\int_{-\infty}^{t}\int_{\mathbb{R}^{n-1}} \tfrac{\langle \nu_s(y),(x-y,x_n-f(y,s))\rangle}{(t-s)} \\ &\quad \cdot W(X-\rho(y,s),t-s)\sqrt{1+|\nabla_y f(y,s)|^2}\,g(y,s)dyds,\end{aligned}$$

when $X\in D$. If $E,F\subset \mathbb{R}^{n+1}$ let $d(E,F)$ be the parabolic distance between E and F defined by

$$(1.12)\quad d(E,F) = \inf\{|X-Y|+|s-t|^{\frac{1}{2}} : (X,t)\in E,\ (Y,s)\in F\}.$$

For $\gamma>0$ and $\rho(x,t)\in\partial D$, we let $\Gamma(x,t)=\Gamma_\gamma(x,t)$ be the parabolic cone with vertex at $\rho(x,t)$ defined by

$$\Gamma(x,t) = \{(Y,s)\in \mathbb{R}^{n+1} : d(\{(Y,s)\},\{\rho(x,t)\}) < (1+\gamma)(y_n - f(x,t))\}.$$

If $h: D\to \mathbb{R}$ define the nontangential maximal function of h for a fixed γ by

$$h^*(x,t) = \sup\{h(Y,s) : (Y,s)\in\Gamma_\gamma(x,t)\cap D\}.$$

Note that $h^*: \partial D\to \mathbb{R}$. By $\lim_{(Y,s)\to\rho(x,t)} h(Y,s)$, we shall always mean the limit as $(Y,s)\to\rho(x,t)$ in $\Gamma(x,t)$. With this notation we prove in section 2

Theorem 5 *Let f satisfy (1.1)-(1.3) and have compact support in $\mathbb{R}^n$. For fixed $p, 1<p<\infty$, there exists $\delta_0=\delta_0(n,p)$, $\hat{c}_0=\hat{c}_0(n,p)$, and $\gamma=\gamma(n)$ so that the following statement is true. If $a_1,a_2\le\delta_0$, then given $g\in L^p(\mathbb{R}^n)$ there corresponds $k\in L^p(\mathbb{R}^n)$ with $\|k\|_p\le \hat{c}_0\|g\|_p$ and such that Lk is the unique solution to the heat equation in D for which (1.13), (1.14) hold :*

$$(1.13)\quad \lim_{(Y,s)\to\rho(x,t)} Lk(Y,s) = g(x,t)$$

for almost every (x,t) *with respect to Lebesgue* n *measure,*

$$\|(Lk)^*\|_p \leq \hat{c}_0 \|g\|_p, \tag{1.14}$$

where the limit and nontangential maximal function are with respect to $\Gamma_\gamma(x,t)$. k *is the solution to the integral equation*

$$\tfrac{1}{2}\, k(x,t) - \mathcal{L}k(x,t) = g(x,t) \tag{1.15}$$

for almost every $(x,t) \in R^n$.

From Theorem 5 we observe for small a_1, a_2 that the L^p Dirichlet problem for D has a unique solution in the above sense. We shall deduce Theorem 5 in a more or less standard way (as in [FR], [B1], [LS]) from Corollary 1 in chapter 2. The same layer potential techniques can be used to solve the Neumann problem for small a_1, a_2. More specifically,

Theorem 6 *Let* f *satisfy (1.1)-(1.3) and have compact support in* R^n . *For fixed* $p, 1 < p < \infty$, *there exists* $\delta_1 = \delta_1(n,p), \hat{c}_1 = \hat{c}_1(n,p)$, *and* $\gamma = \gamma(n)$ *so that the following statement is true. If* $a_1, a_2 \leq \delta_1$, *then given* $g \in L^p(R^n)$ *there corresponds* $k \in L^p(R^n)$ *with* $\|k\|_p \leq \hat{c}_1 \|g\|_p$ *and such that* $Sk + c$, $c \in R$, *are all solutions to the heat equation in* D *for which (1.16), (1.17) hold :*

$$\lim_{(Y,s)\longrightarrow\rho(x,t)} \langle \nabla_Y k(Y,s), \nu_t(x) \rangle = g(x,t) \tag{1.16}$$

for almost every (x,t) *with respect to Lebesgue* n *measure,*

$$\|(\tfrac{\partial Sk}{\partial x_i})^*\|_p \leq \hat{c}_1 \|g\|_p \text{ for } 1 \leq i \leq n, \tag{1.17}$$

where the limit and nontangential maximal functions are with respect to $\Gamma_\gamma(x,t)$. k *is the solution to the integral equation*

$$\tfrac{1}{2}\, k(x,t) - \mathcal{K}k(x,t) = g(x,t) \tag{1.18}$$

for almost every $(x,t) \in R^n$.

We shall not prove Theorem 6 in this paper although its proof is essentially the same as Theorem 5. In fact our main interest in Theorem 5 is its use in proving the following corollary.

Corollary 1 *Let* f *and* (X,t) *be as in Theorem 1. There exists* $\delta_2 > 0$, *such that if* $a_1, a_2 \leq \delta_2$, *then Theorem 1 is true.*

The proof of Corollary 1 is also given in section 2. To prove Theorem 1, we first observe from Corollary 1 of chapter 2 that this theorem is valid when a_1, a_2 are small. We then use a pertubation type argument as in [LM] to deduce the general case from the small case. One of the key ingredients in the proof are some extension lemmas (see sections three and five) in which we extend functions with certain properties on a set $\subset Q^*$ to a function with these properties on Q^*. These extensions are much more difficult to obtain in R^n , $n > 2$, than their corresponding analogues in R . Our extension lemmas along with some comparison theorems of Brown (see section 4) enable us to start the pertubation argument and successively reduce back to the case of small a_1, a_2. To prove Theorem 2 we argue similarly using results from chapter 2 and [JK] but we shall also need a slight generalization of one of Brown's theorems (see Lemma 6.6). Finally in section 7 we make some remarks concerning all three of our papers in this memoir. We would like to thank Russell Brown for many helpful conversations and in particular for showing us a proof of (1.13) for smooth domains.

2. Nontangential Limits.

A key element in the proof of Theorem 5 is Corollary 1 in chapter 2. Recall from this corollary that if $h \in L^p(R^n), 1 < p < \infty$, then

$$\mathcal{L}h(x,t) = \lim_{\epsilon \to 0} \mathcal{L}_\epsilon h(x,t) \tag{2.1}$$

exists for almost every $(x,t) \in R^n$, relative to Lebesgue measure on R^n and for some $c^+ = c^+(a_1, a_2, n, p) > 0$ we have

$$\|\bar{\mathcal{L}}h\|_p \leq c^+ \|h\|_p \tag{2.2}$$

with

$$\lim_{a_1, a_2 \to 0} c^+(a_1, a_2, n, p) = 0. \tag{2.3}$$

As usual, $\bar{\mathcal{L}}$ denotes the maximal operator associated with $\mathcal{L}$ defined by

$$\bar{\mathcal{L}}h(x,t) = \sup_{\epsilon > 0} |\mathcal{L}_\epsilon h(x,t)|.$$

Proof. To prove (1.15) we observe from (2.2), (2.3) and the usual Banach space argument that if $\mathcal{L}^i = \mathcal{L} \circ \mathcal{L} \cdots \circ \mathcal{L}$ (i times), then the Neumann series $\sum_{i=0}^{\infty} 2^{i+1} \mathcal{L}^i g$, converges in $L^p(R^n)$ for sufficiently small a_1, a_2 to k satisfying (1.15) and the desired norm inequalities. Next we establish (1.14) of Theorem 5 for $0 < a_1, a_2, < \infty$, with constants depending only on a_1, a_2, p, n. It will be clear from the proof that these constants can be chosen independent of a_1, a_2 when a_1, a_2 are sufficiently small. Given $k \in L^p(R^n), 1 < p < \infty$, let $M^{(1)}k(\cdot, t)$, $M^{(2)}k(x, \cdot)$, denote Hardy - Littlewood maximal functions of k defined with respect to balls in R^{n-1}, R, containing x, t, while the other variables are held constant. Put

$$\hat{k}(x,t) = [M^{(1)} \circ M^{(2)} + M^{(2)} \circ M^{(1)}]k(x,t)$$

and observe from the Hardy - Littlewood maximal theorem that $\hat{k} \in L^p(R^n)$. Put $\gamma = \frac{1}{100(a_1^2 + a_2^2 + 1)}$ and note from (1.1), (1.4) that $\Gamma_{2\gamma}(x,t) \subset D$ whenever $(x,t) \in R^n$. To prove (1.14) let $(Z, \tau) \in \Gamma_\gamma(x,t)$ with $z_n = f(x,t) + \lambda$, $\lambda > 0$. Then from our choice of γ we see that

$$|z - x| + |\tau - t|^{\frac{1}{2}} \leq \frac{3\lambda}{\sqrt{a_1^2 + a_2^2 + 1}}. \tag{2.4}$$

To prove (1.14) of Theorem 5 it clearly suffices in view of (2.2) and the Hardy - Littlewood maximal theorem to show that

(2.5) $$|Lk(Z,\tau)| \leq c(a_1,a_2,n)[\bar{\mathcal{L}}+\hat{k}](x,t).$$

For this purpose let $K(X,t,y,s)$ denote the kernel in (1.11) defining $Lg(X,t)$. Given $A \geq 2$, we write

(2.6)
$$Lk(Z,\tau) = \int_{\boldsymbol{R}^n} K(Z,\tau,y,s)\,k(y,s)dyds = \sum_{i=0}^{4}\int_{E_i} K(Z,\tau,y,s)k(y,s)dyds$$
$$= T_0 + T_1 + T_2 + T_3,$$

where

$$E_0 = E_0(A\lambda)$$
$$= \{(y,s) : |y-x|^2 + |f(y,s)-f(x,t)|^2 \leq (A\lambda)^2 \text{ and } |s-t| \leq (A\lambda)^2\}$$
$$E_1 = E_1(A\lambda)$$
$$= \{(y,s) : |y-x|^2 + |f(y,s)-f(x,t)|^2 \leq (A\lambda)^2 \text{ and } |s-t| > (A\lambda)^2\}.$$
$$E_2 = E_0(A\lambda)$$
$$= \{(y,s) : |y-x|^2 + |f(y,s)-f(x,t)|^2 > (A\lambda)^2 \text{ and } |s-t| \leq (A\lambda)^2\}$$
$$E_3 = E_3(A\lambda)$$
$$= \{(y,s) : |y-x|^2 + |f(y,s)-f(x,t)|^2 > (A\lambda)^2 \text{ and } |s-t| > (A\lambda)^2\}.$$

Observe that $E_i \cap E_j = \emptyset$ for $i \neq j$, and $\cup E_i = \ R^n$. To estimate T_0 we note from (1.1), (1.4) that

(2.7) $$|K(Z,\tau,y,s)| \leq c\,\frac{|z-y| + |s-\tau|^{\frac{1}{2}}}{|s-\tau|^{n/2+1}}\ \exp[-\frac{\lambda^2}{c|s-\tau|}]$$

whenever $(y,s) \in \ R^n$. Here $c = c(a_1,a_2,n)$. Clearly (2.7) implies

(2.8) $$|K| \leq c(a_1,a_2,n)\,A\lambda^{-(n+1)} \text{ on } E_0.$$

Integrating (2.8) over E_0 and using (2.4) we conclude

(2.9) $$|T_0| \leq c(a_1,a_2,n)\,A^{n+2}\,\hat{k}(x,t).$$

To estimate T_1, T_2 we again use (1.1), (1.4) to deduce that

(2.10) $$|K(Z,\tau,y,s)| \leq c(a_1,a_2,n)\,\frac{|z-y| + |s-\tau|^{\frac{1}{2}}}{|s-\tau|^{n/2+1}}\ \exp\left[-\frac{|y-z|^2}{4|s-\tau|}\right].$$

From (2.10), (2.4) we see for $(y,s) \in E_1$ and sufficiently large $A = A(a_1,a_2,n)$ that

(2.11) $$|K(Z,\tau,y,s)| \leq c(a_1,a_2,n)\,|s-t|^{-(n+1)/2}.$$

Using (2.11) as in (4.28), (4.29) of chapter 2 we deduce that

$$|T_1| \le c(a_1, a_2, n)\, \hat{k}(x,t). \tag{2.12}$$

From (2.10), (2.4), we also find for sufficiently large $A = A(a_1, a_2, n)$ that when $(y,s) \in E_2$

$$|K(Z, \tau, y, s)| \le c(a_1, a_2, n)\, |x - y|^{-(n+1)}. \tag{2.13}$$

Using (2.13) and arguing as in (4.30) of chapter 2 we obtain

$$|T_2| \le c(a_1, a_2, n)\, \hat{k}(x,t). \tag{2.14}$$

To handle T_3 we first observe from (2.4) and the mean value theorem of differential calculus that if $X = (x, f(x,t))$, then for $(y,s) \in E_3$

$$\begin{aligned} |K(Z, \tau, y, s) - K(X, t, y, s)| \le c(a_1, a_2, n)\, \lambda \left(\tfrac{|y-x|^4 + |s-t|^2}{|s-t|^{n/2+3}} \right) \\ \cdot \exp\left[-\tfrac{|y-x|^2}{100|s-t|} \right], \end{aligned} \tag{2.15}$$

for sufficiently large $A = A(a_1, a_2, n)$. Fix $A = A(a_1, a_2, n)$ so that (2.9)-(2.15) are valid. Using (2.15) and estimates similar to the above (see for example the proof of (4.31) in chapter 2), we conclude that

$$|T_3 + \mathcal{L}_{A\lambda}\, k(x,t)| \le c(a_1, a_2, n)\, A^{-1}\, \hat{k}(x,t). \tag{2.16}$$

Putting (2.16), (2.14), (2.12), and (2.9) in (2.6), we deduce that (2.5) is true. Thus (1.14) of Theorem 5 is valid .

To prove (1.13) we note from a theorem of Strichartz [S], as in chapters 1 and 2, that (1.2), (1.3) are equivalent to

$$\int_{J_n} \int_{J_n} \frac{(f(x^*, t) - f(x^*, s))^2}{(s-t)^2}\, ds dt \le c a_2^2 d \tag{2.17}$$

whenever $(x^*, t^*) \in R^n$, $d > 0$, and J_n is as in (1.5). From this equality and (1.4) it is easily shown that

$$\lim_{s \to t} |s - t|^{-\frac{1}{2}}\, |f(x^*, s) - f(x^*, t)| = 0$$

for almost every $(x^*, t) \in R^n$. Also from (1.1) we see that $\nabla_x f(x,t)$ exists for almost every $(x,t) \in R^n$ with respect to Lebesgue n measure. Using these facts and essentially Egoroff's theorem we see for given $\delta > 0$ that there exists a closed set $F \subset R^n$ with $|\, R^n \setminus F| \le \delta$ and the property that if $(x,t) \in F$ and $\epsilon > 0$, then for some $\eta_0 = \eta_0(\epsilon) > 0$ we have

(2.18)

(a) $|f(x,t) - f(x,s)| \le \epsilon |s-t|^{1/2}$ for $(x,t) \in F$ and $|s-t| \le \eta_0^2$,

(b) $|\nabla_x f(x,t) - \nabla_y f(y,s)| \le \epsilon$ for $(x,t), (y,s) \in F$, and $|x-y| + |t-s|^{\frac{1}{2}} \le \eta_0$,

(c) $|f(y,t) - f(x,t) - \langle\, \nabla_x f(x,t), y - x \,\rangle\, | \le \epsilon\, |x-y|, (x,t) \in F, |y-x| \le \eta_0$.

Also from a familiar point of density argument we deduce the existence a closed set $F_1 \subset F$ with $|F \setminus F_1| \leq \delta$ such that if $(x,t) \in F_1$, and $\epsilon > 0$, then for some $\eta_1 = \eta_1(\epsilon) > 0$ we have

$$|Q_d(x,t) \setminus F| \leq \epsilon\, |Q_d(x,t)| \tag{2.19}$$

whenever $0 < d < \eta_1^2$. To prove (1.13) first suppose $k \in C_0^\infty(\ R^n\ )$ and that (x^*,t^*) is a point in F_1 for which $\mathcal{L}1(x^*,t^*)$ exists. We write

$$Lk(Z,\tau) = L(k - k(x^*,t^*))(Z,\tau) + k(x^*,t^*)\,L1(Z,\tau).$$

Here the integral involving 1 is interpreted as a principal value near ∞. Then from symmetry and the smoothness of k we deduce using estimates similar to those in the proof of (1.14) that

$$\lim_{(Z,\tau)\longrightarrow\rho(x^*,t^*)} L(k - k(x^*,t^*))(Z,\tau) = \mathcal{L}(k - k(x^*,t^*))(x^*,t^*) \tag{2.20}$$

where again the latter integral is interpreted as a principal value near ∞.

To estimate $L1(Z,\tau)$ we again write $Z = (z, f(x^*,t^*) + \lambda)$ and put $L1(Z,\tau) = \sum_{i=0}^3 T_i$, where T_i, $0 \leq i \leq 3$, are defined as in (2.6) with $k \equiv 1$. Next for given $\epsilon > o$, let $\eta = \eta(\epsilon) = \min[\eta_0(\epsilon), \eta_1(\epsilon)]$ and suppose λ, A, E_i are as in (2.6) with $0 < A^2\lambda < \eta$ and $0 \leq i \leq 3$. To estimate T_0 we write $K(Z,\tau,y,s) = H(y,s) + e(y,s)$, where

$$\begin{aligned} H(y,s) \;&= \frac{\lambda + \langle \nabla_{x^*} f(x^*,t^*), (x^*-z)\rangle}{2(4\pi)^{n/2}}\, |s-t^*|^{-(n/2+1)}\,\chi_{[0,\infty]}(t^*-s) \\ &\cdot \exp\left[-\frac{|z-y|^2 + |\lambda + \langle \nabla_{x^*} f(x^*,t^*), x^*-y\rangle|^2}{4|s-t^*|}\right] \end{aligned}$$

when $(y,s) \in\ R^n$. From (2.18) and the mean value theorem we deduce that

$$|e(y,s)| \leq c\,\epsilon \left(\frac{|z-y| + \lambda + |s-\tau|^{\frac{1}{2}}}{|s-\tau|^{n/2+1}} \right) \exp[-\frac{\lambda^2}{c|s-\tau|}] \tag{2.21}$$

for $(y,s) \in F \cap E_0(A\lambda)$. Again, $c = c(a_1,a_2,n)$. Using (2.19), (2.21), and arguing as in (2.7) - (2.9), we get

$$|T_0 - \int_{E_0(A\lambda)} H(y,s)\,dyds| \leq c(a_1,a_2,n)\,\epsilon\, A^{n+2}. \tag{2.22}$$

Now

(2.23)

$$\int_{E_0(A\lambda)} H(y,s)dyds = \int_{R^n} H(y,s)dyds - \int_{R^n \setminus E_0(A\lambda)} H(y,s)dyds = I_1 + I_2.$$

Since $\lambda + \langle \nabla_{x^*} f(x^*,t^*), (x^*-z)\rangle > 0$, thanks to (2.4), we can evaluate I_1 directly using the Gamma function to get $I_1 = 1/2$. For I_2 it follows from estimates similar to those in (2.11), (2.13) that

$$|I_2| \leq c(a_1,a_2,n)\,A^{-1}.$$

Using these inequalities for I_1, I_2 in (2.23) we conclude first that

$$|\int_{E_0(A\lambda)} H(y,s)dyds - \frac{1}{2}| \le c(a_1,a_2,n)\,A^{-1}$$

and second from (2.22) that

$$|\,T_0 - \frac{1}{2}\,| \le c(a_1,a_2,n)\,(\,\epsilon\,A^{n+2} + A^{-1}\,). \tag{2.24}$$

To estimate T_1 we first note from (2.18) as in (2.11) that

$$|K(Z,\tau,y,s)| \le c(a_1,a_2,n)\,[\,\epsilon|s-t^*|^{\frac{1}{2}} + \lambda\,]\,|s-t^*|^{-(n+2)/2} \tag{2.25}$$

when $(y,s) \in F \cap E_1(A\lambda)$. Second we set $E_1 = \cup_{i=1}^3 G_i$ where

$$G_1 = E_1(A\lambda) \cap F \cap E_0(A^2\,\lambda)$$

$$G_2 = (E_1(A\lambda) \setminus F) \cap E_0(A^2\lambda)$$

$$G_3 = E_1(A\lambda) \setminus (G_1 \cup G_2)$$

and put $P_i = \int_{G_i} K(Z,\tau,y,s)\,dyds$ for $1 \le i \le 3$. From (2.25) we deduce as in (2.11)

$$|P_1| \le c(a_1,a_2,n)(\,\epsilon + A^{-1}\,). \tag{2.26}$$

As for P_2, using (2.11) and (2.19) with $d = A^2\lambda$, we find

$$|P_2| \le (A\lambda)^{-(n+1)}|G_2| \le c(a_1,a_2,n)\,\epsilon\,A^{n+1}\,. \tag{2.27}$$

Also, $|P_3| \le c(a_1,a_2,n)\,A^{-(n+1)}$, as we see using (2.11). From this inequality, (2.26), and (2.27) we obtain

$$|\,T_1\,| = |\,\sum_{i=1}^{3} P_i\,| \le c(a_1,a_2,n)[\,\epsilon\,A^{n+1} + A^{-1}\,]. \tag{2.28}$$

T_2 is estimated similarly using (2.13). That is write $E_2(A\lambda) = \cup_{i=1}^3 V_i$ where V_i is defined in the same way as G_i with $E_1(A\lambda)$ replaced by $E_2(A\lambda)$. If $(y,s) \in V_1$ we note from (2.18) as in (2.13) that

$$|K(Z,\tau,y,s)| \le c(a_1,a_2,n)\,(\epsilon|x^* - y| + \lambda\,)|x^* - y|^{-(n+2)}.$$

Let $W_i = \sum_{i=1}^3 \int_{V_i} K(Z,\tau,y,s)\,dyds$. Using the above inequality we get

$$|W_1| \le c(a_1,a_2,n)\,[\epsilon + A^{-1}].$$

From (2.13), (2.19), we deduce

$$|W_2| \le c(a_1,a_2,n)\,\epsilon\,A^{n+1}\,.$$

Also from (2.13), we obtain $|W_3| \le c(a_1,a_2,n)\,A^{-(n+1)}$. Adding these estimates together we see that

$$|T_2| \le c(a_1,a_2,n)\,[\,\epsilon\,A^{n+1} + A^{-1}\,]. \tag{2.29}$$

Finally, (2.16) remains true with k replaced by $k(x^*,t^*)\chi_{R^n}$ provided the integral involving this function is interpreted as a principal value near ∞. Combining (2.16), (2.24), (2.29), and (2.28) we conclude that

$$(2.30) \qquad |L1(Z,\tau) - \tfrac{1}{2} + \mathcal{L}_{A\lambda}1(x^*,t^*)| \le c(a_1,a_2,n)[A^{-1}+\epsilon].$$

Taking the limit supremum of the left hand side of (2.30) as $\lambda \to 0$ we obtain that the resulting expression is less than or equal to the right hand side of (2.30). Letting $\epsilon \to 0$, and then $A \to \infty$ in this inequality we find

$$\lim_{(Z,\tau)\to\rho(x^*,t^*)} L1(Z,\tau) = \tfrac{1}{2} - \mathcal{L}1(x^*,t^*).$$

Combining this inequality with (2.20) we see that (1.13) is valid when $(x^*,t^*) \in F_1$. Since $|R^n \setminus F_1| \le 2\delta$ and $\delta > 0$ is arbitrary we conclude that (1.13) holds for $k \in C_0^\infty(R^n)$. The general case $k \in L^p(R^n)$ follows from the smooth case and (1.14) in a well known way (see [LS, p 824]). We omit the details. Thus (1.13) is true. Finally uniqueness of Lk as in Theorem 5 follows from uniqueness in the smooth case, the Phragmen-Lindelöf maximum principle for bounded solutions to the heat equation, and (1.14). We refer the reader to [LS, p 825] when $D \subset R^2$. The argument is essentially unchanged in higher dimensions. The proof of Theorem 5 is now complete. □

To prove Corollary 1 we assume that the origin in R^{n+1} is in ∂D or equivalently that $f(0,0) = 0$. This assumption is permissible since caloric measure is invariant under translations of R^{n+1}. We claim that it suffices to prove Corollary 1 when f has compact support in R^n. Indeed, let $\beta \in C_0^\infty([-2,2])$, with $0 \le \beta \le 1$, $|\beta'| \le 1000$, and $\beta \equiv 1$ on $[-1,1]$. Given $m > 0$ a positive integer put $f_m(x,t) = \beta(\frac{|x|}{m})\,\beta(\frac{t}{m})$, when $(x,t) \in R^n$. It is easily checked that f_m, $m = 1,2,\ldots$, satisfies (1.1) with uniform Lipschitz norm less than or equal to $c(n)a_1$. Also using the equivalence of (2.17) and (1.2)-(1.3), it is easily shown that f_m, $m = 1,2,\ldots$, satisfies (1.2), (1.3) with constants less than or equal to $c(n)a_2$. Next from the maximum principle for the heat equation we observe that for large m depending on t,d,t^*, we have $\omega(X,t,E,D(f)) = \omega(X,t,E,D(f_m))$ whenever E is a Borel subset of Q^*. We conclude from these observations that if we prove Corollary 1 for $f_m, m = 1,2,\ldots$, then it also holds for f. Thus our claim is true and so we assume that f has compact support in R^n. Let E be a compact subset of Q^* with $|E| > 0$ and suppose that O is an open set containing E with $|O| \le 2|E|$. Let $g, 0 \le g \le 1$, be a continuous function with compact support in $O \cap \{(x,t) \in R^n : t - t^* \le \frac{51}{100}d\}$ and $g \equiv 1$ on E. Let $\delta_0, \hat{c}_0$, be defined as in Theorem 5 with $p = 2$. If $a_1, a_2 \le \delta_0$, then from Theorem 5 we see there exists $k \in L^2(R^n)$ satisfying (1.13)-(1.15) relative to g with $\|k\|_2 \le c|E|^{1/2}$. Let $\bar{x}_n = g(x^*,t^*) + 10(a_1+a_2)d^{\frac{1}{2}}$ and $\bar{t} = t^* + \frac{3}{5}d$. If $c = c(a_1,a_2,n)$ is large enough and

$$G = \{(x,\bar{x}_n,t) : t > \bar{t} \text{ and } c(|x-x^*|^2 + |t-\bar{t}|) \le d\},$$

then $G \subset \Gamma_\gamma(x^*,t^*)$, where γ is as in Theorem 5. Next put

$$G' = \{(x,t) : (x,\bar{x}_n,t) \in G\},$$

$$u(x,t) = Lk(x,\bar{x}_n,t) \text{ on } G'.$$

From Harnack' s inequality for the heat equation we note that if $\bar{X} = (x^*, \bar{x}_n)$, then

$$|Lk(\bar{X}, \bar{t})| \le c|u(x,t)| \le c(Lk)^*(x,t),$$

whenever $(x,t) \in G'$. From this equality, the definition of parabolic measure, and (1.4) we get

(2.31)

$$\begin{aligned}\omega(\bar{X}, \bar{t}, E) &\le Lk(\bar{X}, \bar{t}) \le c(a_1, a_2, n)\left[|G'|^{-1}\int_{G'}|u(x,t)|^2\,dxdt\right]^{\frac{1}{2}}\\ &\le c(a_1,a_2,n)\,|Q^*|^{-\frac{1}{2}}\,\|(Lk)^*\|_2 \le c(a_1,a_2,n)[\,|E|/|Q^*|\,]^{\frac{1}{2}}.\end{aligned}$$

We now use Lemma 4.1 in section 4 of this chapter to obtain from (2.31) that for (X,t) as in Corollary 1, we have

$$c^{-1}\,\frac{\omega(X,t,E)}{\omega(X,t,Q^*)} \le \omega(\bar{X},\bar{t},E) \le c\left[\frac{|E|}{|Q^*|}\right]^{\frac{1}{2}} \tag{2.32}$$

where $c = c(n, A)$. From (2.32) we conclude the lefthand inequality in (1.6) with $\beta = \frac{1}{2}$ when E is compact and $|E| > 0$. The general case, E Borel, follows from this case, regularity of ω, and the usual measure approximation arguments. Finally the lefthand side of (1.6) implies the righthand side of this inequality for some $\beta > 0$ (see [C F]). Another proof of the righthand side of (1.6) with $\beta = \frac{1}{2}$ can be given using duality and arguing as in [LS, (5.3)] . The proof of Corollary 1 is now complete. □

3. Extension Lemmas.

In this section we prove several lemmas which will be used throughout the proof of Theorems 1 - 4. They are similar to Lemma 4 in [LM] and Lemma 2.1 in chapter 2. To set the stage for these lemmas given $I \subset R$ an interval let $l(I)$ denote the length of I and let λI for $\lambda > 0$ denote the interval with the same center as I and λ times the length. Let $h : \bar{Q}^* \to R$ and put

$$\|h\|_{\hat{Q}^*} = \sup_J \left[l(J)^{-1}\int_J\int_J \frac{(h(x,t)-h(x,s))^2}{(t-s)^2}\,dsdt\right]^{\frac{1}{2}}$$

where the supremum is taken over all intervals $J \subset \bar{J}_n$ and $x \in \bar{J}_1 \times \cdots \times \bar{J}_{n-1}$. Note that if $\|h\|_{\hat{Q}^*} < \infty$, then h satisfies condition (2.17) on Q^* which globally is equivalent to (1.2), (1.3).

We assume that h satisfies

$$|h(x,t) - h(y,t)| \le a|x-y|, \tag{3.1}$$

$$|h(x,t) - h(x,s)| \le b|s-t|^{\frac{1}{2}}, \tag{3.2}$$

when $(x,t), (y,t), (x,s) \in \bar{Q}^*$. Let $\tilde{h}$ be the extension of h to R^n defined as follows: first define $\tilde{h}$ on $R \times \bar{J}_2 \times \ldots \times \bar{J}_{n-1} \times \bar{J}_n$ by requiring that $\tilde{h}(\cdot, x_2, \ldots, x_{n-1}, t)$ be constant on each component of $R \setminus J_1$ whenever

$$(x_2, \ldots, x_{n-1}, t) \in \bar{J}_2 \times \ldots \times \bar{J}_{n-1} \times \bar{J}_n.$$

Second extend $\tilde{h}$ to $R \times R \times \ldots \times \bar{J}_{n-1} \times \bar{J}_n$ by requiring that $\tilde{h}(x_1, \cdot, x_3, \ldots, x_n, t)$ be constant on each component of $R \setminus J_2$ whenever

$$(x_1, x_3, \ldots, x_n, t) \in R \times \bar{J}_3 \times \ldots \times \bar{J}_{n-1} \times \bar{J}_n.$$

Continuing in this manner we get $\tilde{h}$ defined on R^n. It is easily checked that (3.1), (3.2) remain valid for $\tilde{h}$ on R^n. The gist of the next lemma is that we can replace a function h with good properties on a closed set $E \subset \bar{Q}^*$ by a function h_1 with good properties on $\bar{Q}^*$ which agrees with h on E.

Lemma 3.1 *Let $h : \bar{Q}^* \to R$ satisfy (3.1), (3.2) on Q^* and suppose that*

$$\int_{J_n} \frac{(h(x,t) - h(x,s))^2}{(t-s)^2}\, ds \leq \theta^2 \leq b^2 \tag{3.3}$$

for all $(x,t) \in E$ a closed nonempty subset of $\bar{Q}^*$. Then there exists h_1 a real valued function on $\bar{Q}^*$ with the following properties:

(+) $h = h_1$ on E,

(++) h_1 satisfies (3.1), (3.2) with a, b replaced by a', b' where

$$\max\{\, a',\, b',\, \|h_1\|_{\hat{Q}^*} \} \leq c[\,a + (b\theta)^{\frac{1}{2}}\,]. \tag{3.4}$$

Proof. We first proceed as in the proof of Lemma 4 in [LM]. From (3.2), (3.3), and weak type estimates it is easily shown(see the display preceding (2.8) in [LM]) that

$$|h(x,t) - h(x,s)| \leq c(b\theta)^{\frac{1}{2}}\, |s-t|^{\frac{1}{2}} \tag{3.5}$$

whenever $(x,t) \in E$. Let $\tilde{h}$ be the extension of h to R^n defined as above. If $K, L \subset R^n$ let $d(K,L)$ denote the parabolic distance between the sets K and L defined as in (1.12) with n replaced by $n-1$. As in chapter 2, let $\{Q_i = Q_{\rho_i}(y_i, t_i)\}$ be a family of Whitney rectangles with disjoint interiors, $R^n \setminus E = \cup Q_i$, and the property that

$$c^{-2} d(Q_i, E) \leq \rho_i^{\frac{1}{2}} \leq c^{-1} d(Q_i, E) \tag{3.6}$$

for $i = 1, 2, \ldots$. Next let $0 \leq \psi_1 \in C_0^\infty([-1,1])$ with $\psi_1 > 0$ on $[-\frac{1}{2}, \frac{1}{2}]$ and $\int_{\mathbb{R}} \psi_1 dx = 1$. Put $\psi_2(x) = \psi_1(|x|)$, when $x \in R^{n-1}$. Let

$$\begin{aligned} v(x,t) &= \sum_k \psi_1\left(\tfrac{t-t_k}{\rho_k}\right) \psi_2\left(\tfrac{x-y_k}{\sqrt{2(n-1)\rho_k}}\right), \\ v_i(x,t) &= v(x,t)^{-1}\, \psi_1\left(\tfrac{t-t_i}{\rho_k}\right) \psi_2\left(\tfrac{x-y_i}{\sqrt{2(n-1)\rho_k}}\right). \end{aligned} \tag{3.7}$$

If $c \geq 2$ is large enough in (3.6) we observe that $\{v_i\}$ is a C^∞ partition of unity for $R^n \setminus E$. Put

$$h_1(x,t) = \sum_k \tilde{h}(y_k, t_k) v_k(x,t) \tag{3.8}$$

when $(x,t) \in \bar{Q}^*$ and set $h_1 = h$ on E.

From (3.5), (3.6) we see there exists $(z_k, \tau_k) \in E$ with

$$d(\{(y_k, t_k)\}, \{(z_k, \tau_k)\}) \leq c\rho_k^{\frac{1}{2}}$$

for k = 1, 2, ... and

$$|\tilde{h}(y_k, t_k) - \tilde{h}(z_k, \tau_k)| \leq c_3[(b\theta)^{\frac{1}{2}} + a]\rho_k^{\frac{1}{2}} = \eta\rho_k^{\frac{1}{2}}. \tag{3.9}$$

From (3.9) and the usual Whitney argument it follows that

$$|h_1(x,t) - h_1(y,s)| \leq c\eta\, d(\{(x,t)\}, \{(y,s)\}) \tag{3.10}$$

when $(x,t), (y,s) \in \bar{Q}^*$. Moreover,

$$|\frac{\partial h_1}{\partial t}|(x,t) \leq c\eta\rho_k^{-\frac{1}{2}} \tag{3.11}$$

whenever $(x,t) \in Q_k \in \{Q_i\}$. Using (3.9) - (3.11) we can now repeat the proof of Lemma 2.1 in chaoter 2 from (2.20) on to get

$$\|h_1\|_{\bar{Q}^*}^{\wedge} \leq c\eta,$$

for c_3 large enough. We note that an outline of this proof will be given in the proof of Lemma 3.2 after (3.28) . Here we omit the details. From the above inequality and (3.10) we conclude that Lemma 3.1 is true. □

In order to state the next lemma we shall need some more notation. If $F \subset R^n$, let $T(F) = \{t : (x,t) \in F \text{ for some } x \in R^{n-1}\}$. Let $M_k, k = 0, 1, 2, \ldots$, be the family of closed 'rectangles' of length $2^{-k}(2d)^{\frac{1}{2}}$ in the x_i direction for $1 \leq i \leq n-1$ and length $2^{-2k}d$ in the t direction obtained from bisecting the sides of $\bar{Q}^*$. That is if $Q = I_1 \times I_2 \times \cdots \times I_n \in M_k$, then $I_j \subset \bar{J}_j$, $1 \leq j \leq n-1$, is a closed interval of length $2^{-k}(2d)^{\frac{1}{2}}$, obtained from bisecting $\bar{J}_j$ into 2^k intervals of equal length. Moreover, $I_n \subset \bar{J}_n$ is a closed interval of length $2^{-2k}d$, obtained from bisecting $\bar{J}_n$ into 4^k intervals of equal length. Let $M = \cup_{k=0}^{\infty} M_k$ and let $M'' \subset M' \subset M$ be subcollections of closed ' rectangles ' of M with disjoint interiors. Put $E' = \bar{Q}^* \setminus (\cup_{Q \in M'} Q)$. With this notation we prove

Lemma 3.2 *Let h satisfy (3.1), (3.2) on $\bar{Q}^*$ and*

$$\int_{J_n} \frac{(h(x,t) - h(x,s))^2}{(t-s)^2}\, ds \leq \theta_1^2 \leq b^2 \tag{3.12}$$

whenever $(x,t) \in E'$ while

$$\int_{J_n \setminus 4T(Q)} \left(\int_Q \frac{(h(x,t) - h(x,s))^2}{(t-s)^2}\, dx dt \right) ds \leq \theta_1^2 |Q| \leq b^2 |Q| \tag{3.13}$$

whenever $Q \in M'$. *Then there exists* $h_1, h_2 : \bar{Q}^* \to R$ *such that*

(*i*) $h_1 = h$ *on* E',

(*ii*) h_1 *satisfies (3.1), (3.2), with* a, b, *replaced by* a', b', *where*

$$\max\{a', b', \|h_1\|_{\hat{Q}^*}\} \leq c\,[(a+b)^{\frac{n+2}{n+3}}\, \theta_1^{1/(n+3)} + a] \tag{3.14}$$

(*iii*) h_2 *satisfies (3.1), (3.2) with* a, b *replaced by* a'', b'' *where* $\max\{a'', b''\} \leq c(a + a' + b + b')$,

(*iv*) *If* $\bar{Q}_r(\hat{x}, \hat{t}) \in M''$, *then* $h_2 = h$ *on* $\bar{Q}_{\frac{r}{2}}(\hat{x}, \hat{t})$,

(*v*) $h_2 = h_1$ *on* $Q^* \setminus (\cup_{Q \in M''} Q)$.

Proof. We first construct h_1. To do so let $S_k, 0 \leq k \leq n$, denote the set of all relatively open k dimensional faces of rectangles in M'. Let G denote the set of points in $\cup_{Q \in M'} Q$ with the following property: if $(x,t) \in G$, then there exists $\{Q_i\}$ with $(*)\, Q_i \in M'$ for $i = 1, 2, \ldots$, $(**) \lim_{i \to \infty} d(Q_i, \{(x,t)\}) = 0$, and $(***) \lim_{i \to \infty} |Q_i| = 0$. Let $E = \bar{E}' \cup G$ and note that E is closed. We now define h_1. Put $h_1 = h$ on $S_0 \cup E$. By induction suppose we have defined h_1 on

$$H_l = E \bigcup_{k=0}^{l} (\cup_{F \in S_k} F)$$

for $0 \leq l < n$. Let $\tilde{F} \in S_{l+1}$ and note from the definition of M', E, that if $(x,t) \in \tilde{F} \setminus H_l$, then (x,t) lies in the interior of the union of a finite number of members of M'. Thus there exists $F^* \in S_{l+1}$ of minimal $l+1$ dimensional Hausdorff measure among all faces in S_{l+1} containing (x,t). Since the rectangles in M' were obtained by the bisection method we see that $F^* \subset F$ whenever $F \in S_{l+1}$ contains (x,t). We claim that if $(x,t), (z,\tau) \in \tilde{F} \setminus H_l$ and F^*, F', are minimal $l+1$ dimensional faces relative to $(x,t), (z,\tau)$, as defined above, then either $F^* = F'$ or $F^* \cap F' = \emptyset$. Indeed, suppose for example that $F^* \neq F'$ and

$$F^* \subset \bar{Q}_r(\hat{x}, \hat{t}) = I_1 \times I_2 \times \cdots \times I_n \in M', \tag{3.15}$$

where $(y,s) \in F^*$ if

(α) $y_i = \hat{x}_i + (\frac{r}{2})^{\frac{1}{2}}, 1 \leq i \leq n - l - 1$,

(β) $|y_i - \hat{x}_i| < (\frac{r}{2})^{\frac{1}{2}}$, $n - l - 1 < i \leq n - 1$,

(γ) $|s - \hat{t}| < r/2$.

Since $(z,\tau) \notin E$, it is easily shown that (z,τ) lies in a rectangle of the form

$$Q_\rho(\hat{z}, \hat{\tau}) = \tilde{I}_1 \times \tilde{I}_2 \times \cdots \times \tilde{I}_{n-l-1} \times K_1 \times \cdots \times K_{l+1}$$

in M' where either $\tilde{I}_j \subset I_j$ or vice versa for $1 \leq j \leq n - l - 1$ and each $K_i, 1 \leq i \leq l+1$, is a closed interval. From disjointness of the rectangles in M' and our assumption that $F^* \neq F'$ we find that $K_1 \times \cdots \times K_{l+1}$ and $I_{n-l} \times \cdots \times I_n$ have disjoint interiors in the relative topology induced from R^n. Using this fact and the fact that F' is a minimal face containing (z,τ), we conclude that $F' \cap F^* = \emptyset$. From this example, we see that our claim is true.

From our claim we deduce that $\tilde{F} \setminus H_l$ is the union of a countable number of minimal faces. Thus to complete the definition of h_1 on $\tilde{F}$, it suffices to define h_1 on a minimal face $F^* \subset \tilde{F}$. To do this suppose first that $l = 0$. Then F^* is a line segment connecting two points of $H_0 = E \cup S_0$. We extend h_1 to $\bar{F}^*$ by requiring that h_1 be linear on F^*. If $0 < l < n$, we consider two cases. First consider the case when F^* has a side parallel to the t axis. In this case we proceed as in the proof of Lemma 3.1 and write $F^* = \cup Q_i$, where $\{Q_i\}$ are Whitney $l+1$ dimensional rectangles with disjoint interiors. Moreover, $Q_i, i = 1, \ldots,$ has sidelength $\rho_i^{\frac{1}{2}}$ in the space variables, sidelength ρ_i in the t variable, and

$$c^{-2} d(Q_i, \bar{F}^* \setminus F^*) \leq \rho_i^{\frac{1}{2}} \leq c^{-1} d(Q_i, \bar{F}^* \setminus F^*)$$

for $i = 1, 2, \ldots$. Let $\{v_i\}$ be a partition of unity for F^* defined relative to $\{Q_i\}$ in a way similar to (3.7). We also set

$$h_1(x,t) = \sum h_1(y_i, s_i)\, v_i(x,t), \quad (x,t) \in F^*$$

which is similar to (3.8) except that now $(y_i, s_i) \in \bar{F}^* \setminus F^*$ with

$$c^{-2} d(Q_i, \{(y_i, s_i)\}) \leq \rho_i^{\frac{1}{2}} \leq c^{-1}\, d(Q_i, \{(y_i, s_i)\}). \tag{3.16}$$

From the induction hypothesis we see that h_1 is well defined on $\bar{F}^*$. Next consider the case when F^* does not have any of its sides parallel to the t axis (i.e. all points on F^* have the same t coordinate). In this case we write $F^* = \cup Q_i$ where the $Q_i, i = 1, \ldots,$ are Whitney $l+1$ dimensional cubes with disjoint interiors and side length proportional to their distance from $\bar{F}^* \setminus F^*$. We then proceed as in the first case to define h_1. By induction and our previous remark we get h_1 defined on $\bar{Q}^*$.

Next we define h_2 on $\bar{Q}^*$. First let h_2 be as in (iv), (v) of Lemma 3.2 . Then in order to complete the definition of h_2 we need only define h_2 on $Q_r(\hat{x}, \hat{t}) \setminus \bar{Q}_{\frac{r}{2}}(\hat{x}, \hat{t})$ whenever $Q_r(\hat{x}, \hat{t}) \in M''$. To do this we again use essentially the Whitney extension theorem to define

$$h_2(x,t) = \sum h_2(y_i, s_i)\, v_i(x,t), \ (x,t) \in Q_r(\hat{x}, \hat{t}) \setminus \bar{Q}_{\frac{r}{2}}(\hat{x}, \hat{t}),$$

where $\{v_i\}$ is a partition of unity for $Q_r(\hat{x}, \hat{t}) \setminus \bar{Q}_{\frac{r}{2}}(\hat{x}, \hat{t})$ corresponding to the Whitney rectangles $\{Q_i\}$. Moreover, $(y_i, s_i), i = 1, 2, \ldots,$ are points in $\partial Q_r(\hat{x}, \hat{t}) \cup \bar{Q}_{\frac{r}{2}}(\hat{x}, \hat{t})$ whose parabolic distance to Q_i is proportional to the side length of Q_i in the x direction. This completes the definition of h_2 on Q^*.

We now show that h_1, h_2 satisfy (3.1), (3.2) with the desired constants. For this purpose we note from (3.1), (3.2) for h, (3.12)-(3.13), and Lebesgue dominated convergence that (3.12) holds for h whenever $(x,t) \in E$. From (3.12), (3.1), (3.2), and weak type estimates it follows as in the display preceding (2.8) of [LM] that

$$|h(x,t) - h(x,s)| \leq (b\theta_1)^{\frac{1}{2}}\, |s-t|^{\frac{1}{2}}, \tag{3.17}$$

whenever $(x,t) \in E$. A somewhat more involved argument using (3.13), (3.1), (3.2), also gives

(3.18)

$$\int_{\{s \in J_n : |s-t| \geq 8l(T(Q))\}} \frac{(h(x,t) - h(x,s))^2}{(t-s)^2}\, ds \leq c \left[(a+b)^{2(n+1)}\, \theta_1^4 \right]^{\frac{1}{n+3}}$$

when $(x,t) \in Q \in M'$. Indeed from (3.13) and weak type estimates we see that for each $\lambda > 0$,

$$T = \int_{\{s\in J_n : |s-t|\geq 8l(T(Q))\}} \frac{(h(x,t)-h(x,s))^2}{(t-s)^2}\, ds \leq \lambda \tag{3.19}$$

except for $(x,t) \in L \subset Q$ where $|L| \leq \lambda^{-1}\theta_1^2|Q|$. If $\theta_1^2/\lambda < 1$ and $(x,t) \in L$, we observe that there exists $(x_1,t_1) \in Q \setminus L$ with

$$d(\{(x_1,t_1)\},\{(x,t)\}) \leq c(\lambda^{-1}\theta_1^2|Q|)^{\frac{1}{n+1}} = r.$$

Moreover

$$|h(x_1,t_1) - h(x,t)| \leq c(a+b)r$$

and

$$|h(x,s) - h(x_1,s)| \leq car$$

whenever $(x,s),(x_1,s) \in \bar{Q}^*$. Using these inequalities, (3.19) for (x_1,t_1), and the triangle inequality, we deduce that

$$\begin{aligned} T &\leq \int_{\{s\in J_n : |s-t|\geq 8l(T(Q))\}} \frac{(h(x_1,t_1)-h(x_1,s))^2}{(t-s)^2}\, ds \\ &\quad + c(a+b)^2 r^2 \int_{\{s\in J_n : |s-t|\geq 8l(T(Q))\}} (s-t)^{-2} ds \\ &\leq c\lambda + c(a+b)^2(\lambda^{-1}\theta_1^2)^{\frac{2}{n+1}}. \end{aligned} \tag{3.20}$$

Choosing $\lambda = (a+b)^{\frac{2(n+1)}{n+3}}\,\theta_1^{\frac{4}{n+3}}$ in (3.20) we obtain (3.18) for $(x,t) \in \bar{Q}^*$. (3.18), (3.2) and weak type estimates give

$$|h(x,t) - h(x,s)| \leq c\,(a+b)^{\frac{n+2}{n+3}}\,\theta_1^{1/(n+3)} \max\{|s-t|^{\frac{1}{2}}, l(T(Q))^{\frac{1}{2}}\} \tag{3.21}$$

for $(x,t) \in Q \in M'$. We now use (3.21), (3.17), and induction to show the existence of $c_4 \geq 2$ such that

$$|h_1(x,t) - h_1(y,s)| \leq \xi d(\{(x,t)\},\{(y,s)\}) \tag{3.22}$$

when $(x,t),(y,s) \in Q^*$, where $\xi = c_4[(a+b)^{\frac{n+2}{n+3}}\,\theta_1^{1/(n+3)} + a]$.

It is easily seen that (3.22) holds on H_0 since $h = h_1$ on this set. Next suppose $F^* \subset S_1$ is a minimal line segment . Then from (3.22) for H_0 and linearity of h_1, we deduce that (3.22) holds on F^*. For later use we also observe that h_1 is Lipschitz on F^* with

$$\left|\frac{\partial h_1}{\partial t}\right| \leq c\,\xi\, l(F^*)^{-\frac{1}{2}}. \tag{3.23}$$

Thus (3.22) holds on $\bar{F}^*$. To show (3.22) holds on H_1 we observe that if F^*, F' are distinct minimal line segments with $(x,t) \in F^*, (y,s) \in F'$ then there exists

$(x_1, t_1) \in \bar{F}^* \cap H_0$, and $(y_1, s_1) \in \bar{F}' \cap H_0$, such that

(3.24)
$$\min\{d(\{(x,t)\},\{(x_1,t_1)\}),\ d(\{(y,s)\},\{(y_1,s_1)\}),\ d(\{(y_1,s_1)\},\{(x_1,t_1)\})\} \leq c d(\{(x,t)\},\{(y,s)\})$$

Using (3.24), the triangle inequality, the induction hypothesis, and (3.22) for $(x_1, t_1) \in \bar{F}^*$, $(y_1, s_1) \in \bar{F}'$, we get

$$\begin{aligned} |h_1(x,t) - h_1(y,s)| &\leq |h_1(x,t) - h_1(x_1,t_1)| \\ &+ |h_1(x_1,t_1) - h_1(y_1,s_1)| + |h_1(y,s) - h_1(y_1,s_1)| \\ &\leq \xi\, d(\{(x,t)\},\{(y,s)\}), \end{aligned} \tag{3.25}$$

which is (3.22) for $(x,t) \in F^*$ and $(y,s) \in F'$. If $(x,t) \in F^*$ and $(y,s) \in H_0$, a similar argument applies. Thus (3.22) is valid for h_1 restricted to H_1. By induction, suppose we have proved that (3.22) holds for h_1 restricted to H_l when $0 < l < n$. If $F^* \in S_{l+1}$ is a minimal $l+1$ dimensional face, then from the induction hypothesis and the usual Whitney type argument, we see that (3.22) holds on F^* for c_4 suitably large. If F^*, F', are distinct minimal $l+1$ dimensional subfaces with $(x,t) \in F^*, (y,s) \in F'$, then there exists $(x_1,t_1) \in \bar{F}^* \cap H_l$, $(y_1,s_1) \in \bar{F}' \cap H_l$ for which (3.24) is valid. Using (3.24), the induction hypothesis, and (3.22) for $\bar{F}^*, \bar{F}'$ as in (3.25) we conclude that (3.22) holds when $(x,t) \in F^*, (y,s) \in F'$. A similar argument can be given when $(x,t) \in F^*$ and $(y,s) \in H_l$. We conclude first that (3.22) holds for h_1 restricted to H_{l+1} and thereupon by induction that (3.22) is valid on $\bar{Q}^*$ for h_1. Thus (ii) of Lemma 3.2 is true for a', b'. From (3.22) for h_1, (3.1), (3.2) for h, our definition of h_2, and the usual Whitney argument we conclude that (iii) of Lemma 3.2 is also true for h_2.

It remains to show that

$$\|h_1\|_{\hat{Q}^*} \leq c\xi \tag{3.26}$$

for c_4 suitably large in order to complete the proof of Lemma 3.2 . To prove (3.26) we need to show for an arbitrary closed interval $J \subset \bar{J}_n$ and $x \in \bar{J}_1 \times \cdots \times \bar{J}_{n-1}$ that

$$\int_J \int_J \frac{(h(x,t) - h(x,s))^2}{(t-s)^2}\, ds dt \leq c\xi^2\, l(J). \tag{3.27}$$

To outline the proof we use the argument in Lemma 2.1 of chapter 2 after (2.23) to successively reduce the proof of (3.27) to

$$\begin{aligned} &(a)\ \{x\} \times J \subset Q \in M', \\ &(b)\ \{x\} \times J \subset F^* \text{ a minimal } l+1 \text{ dimensional face, } 0 \leq l \leq n-1, \\ &(c)\ \{x\} \times J \text{ a minimal line segment in } H_1. \end{aligned} \tag{3.28}$$

We note that (3.27) for J as in (3.28)(c) is trivial as we deduce from (3.23).

With this game plan in mind let $\Lambda = (\{x\} \times J) \cap (E \cup S_0)$ and note that Λ is closed. We write,

$$\{x\} \times J = \Lambda \bigcup (\cup I_j)$$

where $I_j \subset J$, $j = 1, 2, \ldots$, are the components of $(\{x\} \times J) \setminus \Lambda$ and so are relatively open intervals with disjoint interiors. Clearly for each $I \in \{I_j\}$ there exists $Q = Q(I) \in M'$ with $\{x\} \times I \subset Q$. Moreover either I contains an endpoint of J or $T(Q) = I$. The first possibility can occur for at most two $I \in \{I_j\}$. Let P be the set of intervals I for which the second possibility occurs.

We assume that (3.27) is true when J is replaced by $I \in \{I_j\}$. Under this assumption we get (3.27) for J as follows: from (3.22) we see as in (2.12) of chapter 2 that

$$\int_{16I \cap J_n} \int_I \frac{(h_1(x,t) - h_1(x,s))^2}{(t-s)^2}\, dsdt \leq c\xi^2\, l(I) \tag{3.29}$$

whenever $I \in \{I_k\}$. Next we note that

(3.30)

$$\int_J \int_J \frac{(h_1(x,t) - h_1(x,s))^2}{(t-s)^2}\, dsdt = \int_\Lambda \int_\Lambda \frac{(h_1(x,t) - h_1(x,s))^2}{(t-s)^2}\, dsdt$$

$$+ 2\sum_k \int_{I_k} \int_\Lambda \frac{(h_1(x,t) - h_1(x,s))^2}{(t-s)^2}\, dsdt + \sum_{k,j} \int_{I_k} \int_{I_j} \frac{(h_1(x,t) - h_1(x,s))^2}{(t-s)^2}\, dsdt$$

$$= T_1 + T_2 + T_3.$$

We note that (3.30) is just (2.22) in chapter 2 with $\tilde{\phi}$ replaced by h_1 and $\alpha = \frac{1}{2}$. Since $h_1 = h$ on Λ we have from (3.12) that

$$T_1 \leq \theta_1^2\, l(J). \tag{3.31}$$

To handle T_2 note from our definition of h_1, P, that if $I_k \in P$ and $(x,t) \in \{x\} \times I_k \subset Q$, then there exists $(z_k, \tau_k) \in H_0$ with

$$c^{-1}\, l(I_k)^{\frac{1}{2}} \leq |x - z_k| + |t - \tau_k|^{\frac{1}{2}} \leq c\, l(I_k)^{\frac{1}{2}}.$$

Since $h_1 = h$ on H_0 we see from this inequality, (3.21), and (3.22) that

$$|h(x,t) - h_1(x,t)| \leq c\xi l(I_k)^{\frac{1}{2}},\ (x,t) \in I_k. \tag{3.32}$$

In view of (3.18), (3.29), $h_1 = h$ on H_0, and (3.32) we see that (2.26) of chapter

2 can be copied with $\tilde{\phi}, \phi_2$, replaced by h_1, h respectively. Doing this we get

(3.33)

$$
\begin{aligned}
T_2 &= 2\sum_k \int_{I_k}\int_\Lambda \frac{(h_1(x,t)-h_1(x,s))^2}{(t-s)^2}\,dsdt \\
&\le c\sum_k l \int_{I_k}\int_\Lambda \frac{(h(x,t)-h(x,s))^2}{(t-s)^2}\,dsdt + c\sum_k \int_{I_k}\int_\Lambda \frac{(h_1(x,t)-h(x,t))^2}{(t-s)^2}\,dsdt \\
&\le c\xi^2 l(J) + \sum_k \int_{I_k}\int_{J_n\setminus 16I_k} \frac{(h(x,t)-h(x,ts))^2}{(t-s)^2}\,dsdt \\
&\qquad + c\sum_k \int_{I_k}\int_{J_n\setminus 16I_k} \xi^2 \tfrac{l(I_k)}{(t-s)^2}\,dsdt \\
&\le c\xi^2 \sum_k l(I_k) \le c\xi^2 l(J).
\end{aligned}
$$

To estimate T_3 we let $P(k) = \{I_j \in P : l(I_j) \le l(I_k)\}$ and copy (2.27) of chapter 2 with $\tilde{\phi} = h_1, \phi_2 = h$ to get

$$
T_3 \le c\xi^2 l(J) + N_1 + N_2 + N_3 \tag{3.34}
$$

where

$$
N_1 = c\sum_{I_k\in P}\sum_{I_j\in P(k)} \int_{I_k}\int_{I_j\setminus(16I_k\cap J_n)} \frac{(h(x,t)-h(x,s))^2}{(t-s)^2}\,dsdt
$$

$$
N_2 = c\sum_{I_k\in P}\sum_{I_j\in P(k)} \int_{I_k}\int_{I_j\setminus(16I_k\cap J_n)} \frac{(h(x,t)-h_1(x,t))^2}{(t-s)^2}\,dsdt
$$

$$
N_3 = c\sum_{I_k\in P}\sum_{I_j\in P(k)} \int_{I_k}\int_{I_j\setminus(16I_k\cap J_n)} \frac{(h(x,s)-h_1(x,s))^2}{(t-s)^2}\,dsdt.
$$

From (3.18) we deduce that

$$
\begin{aligned}
N_1 &\le c\sum_{I_k\in P} \int_{I_k}\int_{J_n\setminus 16I_k} \frac{(h(x,t)-h(x,s))^2}{(t-s)^2}\,dsdt \\
&\le c\xi^2 \sum_{I_k\in P} l(I_k) \le c\xi^2 l(J).
\end{aligned}
$$

From (3.32) we also get

$$
N_2 + N_3 \le c\xi^2 l(J).
$$

Putting these estimates for $N_i, 1 \le i \le 3$, into (3.34) we obtain first that

$$
T_3 \le c\xi^2 l(J)
$$

and thereupon from (3.34), (3.33), (3.31), and (3.30) that (3.27) is valid for arbitrary $J \subset J_n$ provided this inequality holds with J replaced by $I \in \{I_k\}$. Since $\{x\} \times I \in Q(I) \in M'$, we conclude that it suffices to prove (3.27) under assumption (3.28)(a). To prove (3.27) when $\{x\} \times J \subset Q$ we first assume that (3.27) holds when $\{x\} \times J \subset H_l \cap Q$ for fixed $l, 1 \leq l < n$. Under this assumption we show that (3.27) remains valid when $x \in \bar{J}_1 \times \dots \bar{J}_{n-1}$ and $\{x\} \times J \subset H_{l+1} \cap Q$. Indeed, suppose $\{x\} \times J \subset F \in S_{l+1}$. Again we write

$$\{x\} \times J = H \bigcup (\cup I_j)$$

where H is as defined earlier and $I_j \subset \bar{J}_n, j = 1, 2, \dots,$ are disjoint relatively open intervals with the property that if $I \in \{I_j\}$, then there exists $F^* = F^*(I)$, a minimal $l+1$ dimensional face with $\{x\} \times I \subset \bar{F}^*$. Clearly $l(I) = T(F^*)$ except for at most two $I \in \{I_k\}$. Again we claim that if (3.27) is valid with J replaced by $I \in \{I_k\}$ then this inequality also holds for J. The proof follows from (3.29)-(3.34) as previously. We omit the details. Thus to prove that (3.27) holds when $\{x\} \times J \subset F \in S_{l+1}$ it suffices to prove this inequality when $J \subset \bar{F}^*$, a minimal $l+1$ face . If $\{x\} \times J \subset \bar{F}^* \setminus F^*$, then $\{x\} \times J \subset H_l$ so (3.27) holds by assumption. Otherwise the relative interior of $\{x\} \times J$ is a subset of F^* . In this case put

$$\delta = \min\{|x - y| : \{y\} \times J \subset \bar{F}^* \setminus F^* \}.$$

Choose x_1 with $(x_1, t_1) \in \bar{F}^* \setminus F^*$ for some $t_1 \in J$ and $|x - x_1| = \delta$. Then

$$(3.35) \quad \int_J \int_J \frac{(h_1(x,t) - h_1(x,s))^2}{(t-s)^2} \, dsdt = \int_{L_1} \int \dots dsdt + \int_{L_2} \int \dots dsdt$$

where $L_1 = \{(s,t) \in J \times J : |s - t| \geq \delta^2/4\}$ and $L_2 = (J \times J) \setminus L_1$. From our assumption that (3.27) holds on H_l and (3.22), we get

(3.36)
$$\int_{L_1} \int \frac{(h_1(x,t) - h_1(x,s))^2}{(t-s)^2} \, dsdt \leq \int_J \int_J \frac{(h_1(x_1,t) - h_1(x_1,s))^2}{(t-s)^2} \, dsdt$$
$$+ \int_{L_1} \int \frac{\delta^2 \xi^2}{|s-t|^2} \, dsdt$$
$$\leq c\xi^2 l(J).$$

To estimate the integral involving L_2 we recall that

$$h_1(x,t) = \sum h_1(y_i, s_i) v_i(x,t), \; (x,t) \in F^*,$$

where $(y_i, s_i) \in \bar{F}^* \setminus F^*$ and $\{v_i\}$ is a partition of unity for F^* defined relative to the Whitney $l+1$ dimensional rectangles $\{Q_i\}$ satisfying (3.16). Using (3.22) it is easily shown that

$$(3.37) \qquad |\frac{\partial h_1}{\partial t}| \leq c\xi l(T(Q_k))^{-\frac{1}{2}}$$

on $Q_k \in \{Q_i\}$.

Since $\{Q_i\}$ are Whitney rectangles we see the existence of $c_5 \geq 2$ such that

(3.38)
$$\text{card } \{k : Q_k \cap (\{x\} \times J) \neq \emptyset \text{ and } 2^{-j}\delta^2 \leq c_5\, l(T(Q_k)) \leq 2^{-j+1}\delta^2\} \leq c_5$$

for $j = 1, 2, \ldots$, where card denotes cardinality. Now if $Q_k \in \{Q_i\}$ and $l(Q_k) > \delta^2/c_5$, then from (3.22),(3.37) we deduce

$$\int_{T(Q_k)} \int_{\{s \in J_n : |s-t| \leq \frac{\delta^2}{4}\}} \frac{(h_1(x,t) - h_1(x,s))^2}{(t-s)^2}\, ds dt \leq c\xi^2\, l(T(Q_k)). \tag{3.39}$$

If $l(Q_k) \leq \delta^2/c_5$ we use (3.22),(3.37) to obtain

$$\begin{aligned} &\int_{T(Q_k)} \int_{\{s \in J_n : |s-t| \leq \frac{\delta^2}{4}\}} \frac{(h_1(x,t) - h_1(x,s))^2}{(t-s)^2}\, ds dt \\ &\qquad \leq c\xi^2\, l(T(Q_k)) \ln \left[\frac{c\delta^2}{l(T(Q_k))}\right]. \end{aligned} \tag{3.40}$$

Summing (3.39), (3.40) over $Q_k \in \{Q_i\}$ and using (3.38) we conclude that

$$\int\int_{L_2} \frac{(h_1(x,t) - h_1(x,s))^2}{(t-s)^2}\, ds dt \leq c\xi^2 l(J).$$

Using this inequality and (3.36) in (3.35) we get (3.27) first for $\{x\} \times J \subset \bar{F}^*$ and thereupon for $\{x\} \times J \subset H_{l+1}$. To remove the assumption that (3.27) holds for $\{x\} \times J \subset H_l$, we see from downward induction that it suffices to show (3.27) holds when $J \subset H_1$. Moreover, proceeding as above we see that in order to prove this statement it suffices to show that (3.27) is true when $J \subset F^*$, a minimal line segment. (3.27) in this case(as remarked earlier) is easily verified using (3.23). Hence (3.27) is valid whenever $J \subset \bar{J}_n$ is an interval and $x \in \bar{J}_1 \times \cdots \times \bar{J}_{n-1}$. The proof of (3.26) and Lemma 3.2 is now complete □

4. Proof of Theorem 1 in a Special Case.

Let $g : R^n \to R$ satisfy (3.1), (3.2) on R^n with h replaced by g. As in section 1 we denote points in R^n in the space variable by X and write $X = (x, x_n)$, $x \in R^{n-1}$, $x_n \in R$. Also as in section 1 we put $D = D(g) = \{(X,t) : x_n > g(x,t),\, (x,t) \in R^n\}$ and let $\tilde{\omega}, \omega$ be the corresponding measures. Let $Q^* = Q_d(x^*, t^*)$ be as in (1.5) and as in section 2, set

$$\bar{X}(Q^*) = \bar{X}(a, b, Q^*, g) = (x^*,\, g(x^*, t^*) + 10(a+b)d^{\frac{1}{2}})$$

$$\bar{t}(Q^*) = t^* + \tfrac{3}{5}d.$$

We shall need the following lemmas in the proof of Theorem 1.

Lemma 4.1 *Let g satisfy (3.1), (3.2) on R^n and suppose $A \geq 2$. Let $(X,t) \in D$ with $|X - (x^*, g(x^*, t^*))|^2 \leq A(t - t^*)$ and $d \leq t - t^*$. Then there exists $c = c(a, b, A, n) \geq 2$ such that*

$$\omega(X, t, Q^*) \leq c\,\omega(X, t, Q') \tag{4.1}$$

where $Q' = Q_{\frac{d}{2}}(x^*, t^*)$. *Moreover,*

$$(4.2) \qquad c^{-1} \le \omega(\bar{X}, \bar{t}, Q^*),$$

$$(4.3) \qquad c^{-1}\omega(\bar{X}, \bar{t}, E) \le \frac{\omega(X,t,E)}{\omega(X,t,Q^*)} \le c\omega(\bar{X}, \bar{t}, E),$$

whenever E *is a Borel subset of* Q^*.□

Lemma 4.2 *Let* $g, g^* : R^n \to R$ *satisfy (3.1), (3.2) with constants* a, b, *and* a^*, b^*, *respectively. There exists* $\alpha = \alpha(a, a^*, b, b^*, n), 0 < \alpha < 1$ *and* $c = c(a, a^*, b, b^*, n) \ge 2$ *such that if* $g^* = g$ *on a nonempty Borel set* $E \subset Q^*$, *then*

$$(4.4) \qquad c^{-1}\omega(\hat{X}, \hat{t}, E, D(g^*))^{\frac{1}{\alpha}} \le \omega(\tilde{X}, \tilde{t}, E, D(g)) \le c\omega(\hat{X}, \hat{t}, E, D(g^*))^{\alpha}.$$

Here $\hat{X} = \bar{X}(a+a^*, b+b^*, Q^*, g^*)$, $\tilde{X} = \bar{X}(a+a^*, b+b^*, Q^*, g)$, *and* $\hat{t} = \tilde{t} = \bar{t}(Q^*)$.□

Somewhat weaker versions of Lemma 4.1 were proved by [K] and [W]. Lemma 4.1, as above, is in [B2, Lemma 2.2]. Lemma 4.2 can be proved by comparing both parabolic measures for E to parabolic measure for a certain " parabolic sawtooth. " The argument in [B2, Lemma 2.10] can then be repeated to get Lemma 4.2. We note that Lemma 4.1 allows us to compare parabolic measures defined relative to different points in a given graph domain while Lemma 4.2 allows us to compare certain parabolic measures of a set common to the boundary of two different graph domains. Next we prove

Lemma 4.3 *Let* g *satisfy (3.1), (3.2) on* R^n. *There exists* $\theta_2 = \theta_2(n)$ *and* $\theta_i = \theta_i(b, n), 0 < \theta_i < 1, i = 3, 4$, *such that if* $a \le \theta_2$ *and*

$$(4.5) \qquad \int_{J_n} \left(\int_{\bar{Q}^*} \frac{(g(x,t) - g(x,s))^2}{(s-t)^2} dx dt \right) ds \le \theta_3^2 \, |Q^*|,$$

then whenever F *is a Borel subset of* Q^* *with* $|F|/|Q^*| \ge \frac{7}{8}$, *we have*

$$(4.6) \qquad \omega(\bar{X}, \bar{t}, F) \ge \theta_4.$$

Proof. Let h be the restriction of g to $\bar{Q}^*$. We note from (4.5) and weak type estimates that there exists E a closed subset of Q^* such that $|E| \ge \frac{1}{4}|Q^*|$ and

$$(4.7) \qquad \int_{J_n} \frac{(h(x,t) - h(x,s))^2}{(s-t)^2} ds \le 16\theta_3^2$$

whenever $(x,t) \in E$. From (4.7) and the hypotheses of Lemma 4.3 we see for θ_3 small enough that we can apply Lemma 3.1 to get h_1 satisfying (3.4) on Q^*. Let $\tilde{h}_1$ be the extension of h_1 to R^n defined as in section 3. From (3.4) it is easily shown that (3.1), (3.2) hold for $\tilde{h}_1$ on R^n. Moreover, it follows from (3.4) as in [LM, (2.5)] that

$$(4.8) \qquad \|\tilde{h}_1\|^{\wedge}_{R^n} \leq \eta$$

for $\eta = c_6[a+(b\theta_3))^{\frac{1}{2}}]$ and $c_6 \geq 2$ large enough. We note, as in (2.7) of section 2, that (4.8) for $\tilde{h}_1$ is equivalent to the existence of $k : R^n \to R$ with $h(x,\cdot) = I_{\frac{1}{2}} * k(x,\cdot)$ and

$$(4.9) \qquad \|k(x,\cdot)\|_* \leq c\eta$$

whenever $x \in R^{n-1}$. As in section 1, $\|\cdot\|_*$ denotes the BMO norm on R. From (4.9), we see that if θ_2, θ_3 are sufficiently small, then we can apply Corollary 1 with h replaced by h_1. Doing this, using (4.2), and the fact that $|E \cap F|/|Q^*| \geq \frac{1}{8}$, (whenever $|F| \geq \frac{7}{8}|Q^*|$), we get

$$(4.10) \qquad \omega(\hat{X},\hat{t},E\cap F, D(\tilde{h}_1)) \geq c^{-1}$$

provided $\theta_2, \theta_3 > 0$ are small enough where $c = c(b,n) \geq 2$,

$$\hat{X} = \bar{X}(a+a', b+b', Q^*, \tilde{h}_1) \text{ and } \hat{t} = \bar{t}(Q^*).$$

From (4.10), (4.4) of Lemma 4.2 with $g^* = \tilde{h}_1$, (4.1)-(4.3) of Lemma 4.1, we deduce for some $c = c(b,n) \geq 2$ that

$$\omega(\bar{X},\bar{t},F\cap E, D(g)) \geq c^{-1}\omega(\tilde{X},\tilde{t},F\cap E, D(g))$$

$$\geq c^{-2}\omega(\hat{X},\hat{t},F\cap E, D(\tilde{h}_1))^{\frac{1}{\alpha}} \geq c^{-3} = \theta_4.$$

Hence Lemma 4.3 is valid. □

To continue the proof of Theorem 1 we now remove the smallness assumption in (4.5). We prove

Lemma 4.4 *Let $g : R^n \to R$ satisfy (3.1), (3.2) and suppose that*

$$(4.11) \qquad \int_{J_n}\left(\int_{Q^*} \frac{(g(x,t)-g(x,s))^2}{(s-t)^2} dx dt\right) ds \leq \beta^2|Q^*|$$

where $0 < \beta \leq b$. There exists $c_7 \geq 2$ (depending only on n) and $\delta = \delta(\beta,b,n), \kappa = \kappa(\beta,b,n), 0 < \delta, \kappa < 1$, such that if $a \leq \theta_2/c_7$, then whenever $F \subset Q^*$ with $\frac{|F|}{|Q^*|} \geq 1-\delta$, we have $\omega(\bar{X},\bar{t},F) \geq \kappa$.

We note that Lemma 4.4 implies Theorem 1 whenever $a_1 \leq \theta_2/c_7$ and $0 < a_2 < \infty$. Indeed in this situation we can apply Lemma 4.4 with $f = g$ whenever $Q_d(x^*,t^*) \subset R^n$ thanks to the equivalence in (4.8), (4.9). Using (4.1)-(4.3) of Lemma 4.1, it then follows that $\omega(X,t,\cdot,D(f))$ is a doubling measure which is also A_∞ on 'rectangles.' The existence of γ in Theorem 1 now follows from a theorem of Coifman and Fefferman[CF] which characterizes doubling A_∞ weights. Actually these authors prove their theorem for A_∞ defined relative to cubes, but the proof is unchanged if cubes are replaced by 'rectangles.'

Arguments similar to the above note will be used several times in the proof of Lemma 4.4.

Proof. We prove Lemma 4.4 by an inductive type argument on β. Observe from Lemma 4.3 that Lemma 4.4 is valid for fixed $b, 0 < b < \infty$, whenever $0 < \beta \leq \theta_3(b,n)$ with $\delta(\beta,b,n) = \frac{1}{8}$ and $\kappa(\beta,b,n) = \theta_4(b,n)$. Suppose by way of induction that Lemma 4.4 is valid whenever $\theta_3 \leq \beta \leq \beta_1$. We put

$$\epsilon = \left(\frac{\theta_2(n)\,\theta_3(b,n)}{c_8\sqrt{b}}\right)^{10n} \tag{4.12}$$

and shall show for $c_8 = c_8(n) \geq 2$, sufficiently large, that Lemma 4.4 is valid for $\theta_3 \leq \beta \leq \beta_2$ when $\beta_2^2 = \min(b^2, (1+\epsilon^2)\beta_1^2)$, provided $\delta(\beta,b,n)$ and $\kappa(\beta,b,n)$ are defined suitably for $\beta_1 \leq \beta \leq \beta_2$. By induction it then follows that Lemma 4.4 is true.

Let $M_k, k = 0,1,\ldots,\ M = \cup M_k$, be the family of dyadic rectangles defined as in section 3 relative to Q^*. First suppose for some $\tilde{Q} \in M_1$ that

$$\int_{J_n\setminus T(\tilde{Q})}\left(\int_{\tilde{Q}} \frac{(g(x,t)-g(x,s))^2}{(s-t)^2}dx\,dt\right)ds \geq 2^{n+2}\,\epsilon^2\,\beta_1^2\,|\tilde{Q}|. \tag{4.13}$$

Then from (4.11) and (4.13) we deduce

$$\begin{aligned}|Q^*|(1+\epsilon^2)\beta_1^2 &\geq \int_{J_n}\left(\int_{Q^*}\frac{(g(x,t)-g(x,s))^2}{(s-t)^2}dx\,dt\right)ds\\ &\geq \sum_{Q\in M_1}\int_{T(Q)}\left(\int_Q \frac{(g(x,t)-g(x,s))^2}{(s-t)^2}dx\,dt\right)ds + 2^{n+2}\,\epsilon^2\,\beta_1^2\,|\tilde{Q}|.\end{aligned}$$

Since M_1 has cardinality 2^{n+1} and $\frac{|\tilde{Q}|}{|Q^*|} = 2^{-(n+1)}$, it follows from the above equality that for some $Q^+ \in M_1$, we have

$$\int_{T(Q^+)}\left(\int_{Q^+}\frac{(g(x,t)-g(x,s))^2}{(s-t)^2}dx\,dt\right)ds \leq (1-\epsilon^2)\beta_1^2\,|Q^+|. \tag{4.14}$$

In this case we take $\delta(\beta,b,n) = 2^{-(n+1)}\delta(\beta_1,b,n)$ for $\beta_1 \leq \beta \leq \beta_2$. Then from (4.14) and the induction assumption with Q^+ playing the role of Q^* we deduce that the conclusion of Lemma 4.4 is valid with $\omega(\bar{X},\bar{t},F)$ replaced by $\omega(X^+,t^+,F\cap Q^+)$, where $X^+ = X(Q^+), t^+ = t(Q^+)$. Using (4.1)-(4.3) of Lemma 4.1 with Q^* replaced by Q^+, we conclude that $\omega(\bar{X},\bar{t},F\cap Q^+) \geq \kappa$. Hence in this case Lemma 4.4 holds for $\kappa(\beta,b,n)$ small enough. Therefore we assume that (4.13) is false for all $\tilde{Q} \in M_1$. Let $M' \subset M$ be the collection of maximal closed rectangles for which (4.13) holds whenever $\tilde{Q} \in M'$. Observe that

$$\int_{J_n\setminus 4T(\tilde{Q})}\left(\int_{\tilde{Q}}\frac{(g(x,t)-g(x,s))^2}{(s-t)^2}dx\,dt\right)ds \leq 2^{2n+3}\,\epsilon^2\beta_1^2\,|\tilde{Q}| \tag{4.15}$$

since (4.13) is false for the father of $\tilde{Q}$. As in section 3 we put $E' = \bar{Q^*}\setminus(\cup_{Q\in M'}Q)$ and note that (4.13) is false for each $\tilde{Q} \in M$ whenever $\tilde{Q}\cap E' \neq \emptyset$. Using (3.2) for g and dominated convergence we see from this note that

$$\int_{J_n}\frac{(g(x,t)-g(x,s))^2}{(s-t)^2}ds \leq 2^{n+2}\epsilon^2\beta_1^2 \tag{4.16}$$

whenever $(x,t) \in E'$. Next suppose that $\hat{M}$ is a finite subcollection of M' and L a closed subset of E' with

$$\left(\sum_{Q_k \in \hat{M}} |Q_k|\right) + |L| \geq \epsilon^2 |Q^*|. \tag{4.17}$$

We claim that

$$\omega\left[\bar{X}, \bar{t},\, L \bigcup (\cup_{Q_k \in \hat{M}} Q_k)\right] \geq \nu > 0 \tag{4.18}$$

for some $\nu = \nu(b,n)$. To prove this claim we consider two cases. First suppose that $|L| \geq \frac{1}{2}\epsilon^2$. In this case we note from (4.16) that we can apply Lemma 3.1 with h equal to the restriction of g to $\bar{Q}^*$ and $E = L$. We get h_1 satisfying

$$\max\{a', b', \|h_1\|_{\hat{Q}^*}\} \leq c[\, a + (b\beta_1\epsilon)^{\frac{1}{2}}\,] \tag{4.19}$$

We now repeat the argument in the note after the statement of Lemma 4.4. From (4.19) we first deduce (4.8) and then (4.9) with θ_3 replaced by $\beta_1\,\epsilon$. From (4.9) and $|L| \geq \frac{\epsilon^2}{2}$ we see for c_7, c_8 large enough that Corollary 1 can be used to get (4.10) for $\tilde{h}_1$ with $E \cap F$ replaced by L. Using (4.10), (4.3), and Lemma 4.2 once again we conclude that (4.18) is valid in this case. If $|L| < \epsilon^2/2$ we use a well known covering argument to choose a subcollection M'' of $\hat{M}$ with the properties: (+) if $\bar{Q}_r(y,s) \in M''$, then $Q_{2r}(y,s) \cap Q = \emptyset$ whenever $Q \in M''$, and (++) $c\sum_{Q \in M''} |Q| \geq \epsilon^2$ for some $c = c(n)$. We again let h be the restriction of g to $\bar{Q}^*$ and note from (4.15), (4.16) that h, E', M', M'' all satisfy the hypotheses of Lemma 3.2. Applying this lemma we obtain h_1, h_2 on $\bar{Q}^*$ satisfying (i)-(v). Again let $\tilde{h}_1, \tilde{h}_2$ denote the extension of these functions to R^n defined as in section 3. On the one hand if $M^* = \{\bar{Q}_{\frac{r}{2}}(y,s) : \bar{Q}_r(y,s) \in M''\}$, then from Lemma 4.2 with $\tilde{h}_2 = g^*, E = \cup_{Q \in M^*} Q$, we deduce

$$\omega(\hat{X}, \hat{t}, \cup_{Q \in M^*} Q, D(\tilde{h}_2))^{\frac{1}{\alpha}} \leq c\,\omega(\tilde{X}, \tilde{t}, \cup_{Q \in M^*} Q, D(g)) \tag{4.20}$$

where $\alpha(a, a'', b, b'', n)$, $(\hat{X}, \hat{t})$, $(\tilde{X}, \tilde{t})$ are defined as in Lemma 4.2 relative to $g^* = \tilde{h}_2$ and g. On the other hand if $M^+ = \{Q_{2r}(y,s); \bar{Q}_r(y,s) \in M''\}$, then from (4.1) and Lemma 4.2 with $\tilde{h}_2 = g^*, \tilde{h}_1 = g$, $E = (\cup_{Q \in M^+} Q) \setminus (\cup_{Q \in M''} Q)$, we find

$$\begin{aligned} \omega(\tilde{X}, \tilde{t}, E, D(\tilde{h}_1))^{\frac{1}{\alpha_1}} &\leq c\,\omega(\hat{X}, \hat{t}, E, D(\tilde{h}_2)) \\ &\leq c\,\omega(\hat{X}, \hat{t}, \cup_{Q \in M^+} Q, D(\tilde{h}_2)) \leq c^2\,\omega(\hat{X}, \hat{t}, \cup_{Q \in M^*} Q, D(\tilde{h}_2)) \end{aligned} \tag{4.21}$$

where now $\alpha_1 = \alpha_1(a', a'', b', b'', n), (\hat{X}, \hat{t}), (\tilde{X}, \tilde{t})$ are as in Lemma 4.2 with $\tilde{h}_2 = g^*$ and $\tilde{h}_1 = g$. We note from $\max\{a', a'', b', b''\} \leq c(a+b)$ and (4.3) that (4.20) remains true with $(\tilde{X}, \tilde{t})$ replaced by $(\bar{X}, \bar{t})$. Combining the resulting inequality with (4.21) we see that in order to prove claim (4.18) in the case we are considering, it suffices to show

$$\omega(\tilde{X}, \tilde{t}, E, D(\tilde{h}_1)) \geq \nu_1 \tag{4.22}$$

where E is as in (4.21) and $\nu_1 = \nu_1(b,n)$. To do this we note first from (3.14) that

$$\max\{a', b', \|h_1\|_{\hat{Q}^*}\} \leq \xi = c[\,(a+b)^{\frac{n+2}{n+3}}\,(\beta_1\epsilon)^{\frac{1}{n+3}} + a]. \tag{4.23}$$

From (4.23) and $|E| \geq c^{-1}\epsilon^2$, we see for c_7, c_8 large enough that we can again apply the argument in the note after the statement of Lemma 4.4 to get (4.8) and (4.9) with η replaced by ξ. Thus we can again apply Corollary 1 and (4.3) to get first (4.22) and then our claim. Hence in either case claim (4.18) is true.

Finally we use (4.18) to prove Lemma 4.4 for $\beta_1 \leq \beta \leq \beta_2$. To do so we note from (4.13) that

$$\begin{aligned}
&2^{n+2}\,\epsilon^2\,\beta_1^2 \sum_{Q\in M'} |Q| + \sum_{Q\in M'} \int_{T(Q)} \left(\iint_Q \frac{(g(x,t)-g(x,s))^2}{(s-t)^2} dx dt \right) ds \\
&\leq \int_{J_n} \left(\iint_{Q^*} \frac{(g(x,t)-g(x,s))^2}{(s-t)^2} dx dt \right) ds \leq (1+\epsilon^2)\beta_1^2\,|Q^*|
\end{aligned} \tag{4.24}$$

Since $Q^* = E' \bigcup (\cup_{Q\in M'} Q)$, it follows from (4.24) that for $0 < \epsilon < \frac{1}{100}$, either $|E'| > 2\epsilon^2|Q^*|$ or

$$\sum_{Q\in M'} \int_{T(Q)} \left(\iint_Q \frac{(g(x,t)-g(x,s))^2}{(s-t)^2} dx dt \right) ds \leq (1 - 2^{n+1}\epsilon^2)\beta_1^2\,|Q^*|. \tag{4.25}$$

Let N denote the set of $Q \in M'$ such that

$$\int_{J_n\setminus T(Q)} \left(\iint_Q \frac{(g(x,t)-g(x,s))^2}{(s-t)^2} dx dt \right) ds \leq \beta_1^2\,|Q|.$$

If (4.25) holds then clearly

$$\sum_{Q\in M'\setminus N} |Q| \leq (1 - 2^{n+1}\epsilon^2)|Q^*|.$$

Hence if $|E'| \leq 2\epsilon^2|Q^*|$, then

$$\sum_{Q\in N} |Q| \geq 2(2^n - 1)\,\epsilon^2\,|Q^*|\,. \tag{4.26}$$

Let $\delta(\beta, b, n) = \epsilon^2\delta(\beta_1, b, n)$ for $\beta_1 \leq \beta \leq \beta_2$ and let $F \subset Q^*$ satisfy $\frac{|F|}{|Q^*|} \geq 1 - \delta$. If $|E'| > 2\epsilon^2|Q^*|$, we easily deduce the existence of a closed set $L \subset F \cap E'$ with $|L| \geq \epsilon^2|Q^*|$. Applying (4.18) with $\hat{M} = \emptyset$ we see that Lemma 4.4 is valid in this case. If (4.25) holds let N' be the subcollection of N for which

$$|F \cap Q| \geq (1 - \delta(\beta_1, b, n))|Q|. \tag{4.27}$$

whenever $Q \in N'$. From (4.26), (4.27), and our choice of δ we find that

$$\begin{aligned}
&\epsilon^2\delta(\beta_1, b, n)\,|Q^*| \geq |Q^* \setminus F| \geq \sum_{Q\in N\setminus N'} |Q \setminus F| \\
&\geq \delta(\beta_1, b, n) \sum_{Q\in N\setminus N'} |Q| \geq \delta(\beta_1, b, n) \left(2(2^n - 1)\epsilon^2|Q^*| - \sum_{Q\in N'} |Q| \right).
\end{aligned}$$

From this inequality we find the existence of a finite subcollection $\hat{M}$ of N' with

$$\sum_{Q \in \hat{M}} |Q| > \epsilon^2 |Q^*|.$$

Thus (4.17) is true with $L = \emptyset$. From (4.27) we see that the induction hypothesis can be applied to g restricted to each $Q \in \hat{M}$ with F replaced by $F \cap Q$. Doing this and using (4.3) of Lemma 4.1 we deduce for some $c = c(b, n)$ that

$$\begin{aligned} & c\omega(\bar{X}(Q^*), \bar{t}(Q^*), F \cap Q, D(g)) \\ (4.28) \qquad & \geq \omega(\bar{X}(Q^*), \bar{t}(Q^*), Q, D(g)) \cdot \omega(\bar{X}(Q), \bar{t}(Q), F \cap Q, D(g)) \\ & \geq \kappa(\beta_1, b, n)\, \omega(\bar{X}(Q^*), \bar{t}(Q^*), Q, D(g)) \end{aligned}$$

whenever $Q \in \hat{M}$. Summing (4.28) over $Q \in \hat{M}$ and using (4.18), we conclude that

$$\begin{aligned} \omega(\bar{X}(Q^*), \bar{t}(Q^*), F, D(g)) & \geq \kappa(\beta_1, b, n)\, \omega(\bar{X}(Q^*), \bar{t}(Q^*), \cup_{Q \in \hat{M}} Q, D(g)) \\ & \geq \kappa(\beta_1, b, n)\, \nu(b, n) = \kappa(\beta, b, n). \end{aligned}$$

The proof of Lemma 4.4 is now complete. □

5. Proof of Theorem 1.

As noted in section 4, Lemma 4.4 implies that Theorem 1 is valid when $0 < a_1 \leq \theta_2/c_7$ and $0 < a_2 < \infty$. In this section we remove the smallness assumption on a_1 to complete the proof of Theorem 1. To do so we shall need Lemma 3.1 in chapter 2. For the reader's convenience we state this lemma again in Lemma 5.1.

Lemma 5.1 *Let $k : R^n \to R$ satisfy (1.1)-(1.3) on R^n and suppose*

$$\mu = \left\| \frac{\partial k}{\partial x_l} \right\|_\infty = \max_{1 \leq j \leq n-1} \left\| \frac{\partial k}{\partial x_j} \right\|_\infty .$$

Then there exists a closed set $K \subset \bar{Q}_d(x^*, t^*)$ and functions $k_1, k_2 : R^n \to R$ satisfying (1.1) with a_1 replaced by ca_1, such that

$$k_1 = k \text{ on } K, \tag{5.1}$$

$$|K| \geq \tfrac{1}{12} |\bar{Q}_d(x^*, t^*)|. \tag{5.2}$$

Moreover, k_1 is defined by

$$k_1(x, t) = -\sin\phi\, \hat{x}_l + \cos\phi\, k_2(\hat{x}, t) \tag{5.3}$$

where $(\hat{x}, t) \in R^n$,

$$\phi = \pm \frac{1}{2} \left[\tan^{-1}\left(\frac{101\mu}{100}\right) - \tan^{-1}\left(\frac{91\mu}{100}\right)\right]$$

and x differs from $\hat{x}$ in a single coordinate, expressed as a function of $(\hat{x},t)$ by

$$(5.4)\qquad \begin{aligned} x_l(\hat{x},t) &= \cos\phi\,\hat{x}_l + \sin\phi\, k_2(\hat{x},t),\\ x_j &= \hat{x}_j\,, 1 \le j \le n-1,\, j \ne l\,. \end{aligned}$$

The function k_2 satisfies

$$(5.5)\qquad \| \frac{\partial k_2}{\partial \hat{x}_l} \|_\infty \le \left(\frac{24}{25} + \frac{1}{50n} \right)\mu\,,$$

$$(5.6)\qquad \| \frac{\partial k_2}{\partial \hat{x}_j} \|_\infty \le \| \frac{\partial k}{\partial x_j} \|_\infty + \frac{\mu}{50n}\,,\ 1 \le j \le n-1,\ j \ne l.$$

Finally, k_1 satisfies (1.4), and k_2 satisfies (1.2) - (1.4) with a_2, b replaced by $\tilde{a}_2, \tilde{b}$, where

$$\tilde{a}_2 = c\,\hat{a}_2\,[\log(2 + \hat{a}_2)]^{\frac{1}{2\alpha}},$$

$$\hat{a}_2 = \max\{ a_1,\, a_2,\, a_2^{\frac{1}{\alpha^2}}\, \mu^{1-\frac{1}{\alpha^2}} \}$$

We note from Lemma 5.1 that $k = k_1$ on a significant portion (K) of $\bar{Q}_d(x^*,t^*)$. Also, if $|||\nabla_{\hat{x}}\, k_2||| = \sum_1^{n-1} \|\frac{\partial k_2}{\partial \hat{x}_j}\|_\infty$, then (see (3.6) of chapter 2)

$$(5.7)\qquad |||\nabla_{\hat{x}}\, k_2||| \le [1 - \frac{1}{50(n-1)}]|||\nabla_x k|||$$

thanks to (5.5), (5.6).

Proof. Our note and (5.7) enable us to again use an induction type argument to get Theorem 1. More specifically, we induct on $|||\nabla_x f|||$. From Lemma 4.4 we see that Theorem 1 is valid whenever f satisfies (1.1)-(1.3) and in some orthogonal coordinate system $(x_1, \ldots, x_n)$ we have $0 < |||\nabla_x f||| \le \theta_2/c_7$, while $0 < a_2 < \infty$. Suppose by way of induction that Theorem 1 is valid whenever f satisfies (1.1)-(1.3) in some orthogonal coordinate system with, $\theta_2/c_7 \le |||\nabla_x f||| \le \tau$ and $0 < a_2 < \infty$. Under this assumption we show that Theorem 1 is valid whenever $0 < a_2 < \infty$ and

$$(5.8)\qquad \theta_2/c_7 \le |||\nabla_x f||| \le \left(1 - \frac{1}{50(n-1)}\right)^{-1} \tau.$$

Since $a_1 \le |||\nabla_x f||| \le \sqrt{n-1}\, a_1$, it then follows from induction that Theorem 1 is true. To prove (5.8) let $\bar{Q}_d(x^*,t^*) \subset R^n$, put $k = f$ in Lemma 5.1, and assume without loss of generality that $l = 1$ in this lemma. Next for ϕ as in Lemma 5.1 consider the transformation $(\hat{X},t) \to (X,t)$ of R^{n+1} onto R^{n+1} defined by

$$(5.9)\qquad \begin{aligned} x_1 &= \cos\phi\,\hat{x}_1 + \sin\phi\,\hat{x}_n\,,\\ x_j &= \hat{x}_j\,,\ 1 \le j \le n-1\,,\\ x_n &= -\sin\phi\,\hat{x}_1 + \cos\phi\,\hat{x}_n\,. \end{aligned}$$

Clearly this transformation is just a rotation in the space variable. If k_1, k_2 are as in Lemma 5.1, then from (5.3), (5.4), we see that this transformation maps $D(k_2)$ expressed in the $(\hat{X}, t)$ coordinate system 1-1 and onto $D(k_1)$ expressed in the (X, t) coordinate system. Moreover, the induced mapping

$$(\hat{x}, t) \to (\hat{x}, k_2(\hat{x}, t), t) \to (x, k_1(x, t), t) \to (x, t)$$

is a 1 - 1 and onto mapping of R^n onto R^n. If (x^*, t^*) corresponds to (x', t^*) under this mapping we observe that

(5.10)
$$|x - x^*|^2 + |k_1(x, t) - k_1(x^*, t^*)|^2 = |\hat{x} - x'|^2 + |k_2(\hat{x}, t) - k_2(x', t^*)|^2 .$$

From (5.10) and Lemma 5.1 we deduce that if P is mapped onto Q^* under the above mapping, then

(5.11) $$P \subset Q_r(\hat{x}, t^*) = \hat{Q}$$

for $r = c(\tau + \tilde{a}_2)d$ provided $c = c(n)$ is large enough. Here $\tilde{a}_2$ is as in Lemma 5.1.

Let $F \subset Q^*$ with $|F| \geq \frac{23}{24}|Q^*|$. From (5.2) of Lemma 5.1 we see that

(5.12) $$|K_1| = |K \cap F| \geq \tfrac{1}{24}|Q^*|.$$

Next choose K_2 so that K_1 corresponds to K_2 under the mapping $(\hat{x}, t) \to (x, t)$. Using (5.2)-(5.6), (5.10) as in (3.21) of chapter 2 and (5.12), we deduce the existence of $c = c(n)$ such that

$$c\,|K_2| \geq (1 + \tau)^{-1} \int_{K_1} \sqrt{1 + |\nabla_x f_1|^2}(x, t)\, dx dt$$

(5.13)
$$\geq \frac{|K_1|}{1 + \tau} \geq \frac{|Q^*|}{24(1 + \tau)}.$$

Next observe from (5.7) and the induction hypothesis that Theorem 1 is valid with f replaced by k_2 and x by $\hat{x}$. Since $K_2 \subset \hat{Q}$ by (5.11), we see that Theorem 1 can be applied with $k_2 = f$, $E = K_2$, and Q^* replaced by $\hat{Q}$. Using (5.13) and (4.1)-(4.3) of Lemma 4.1 we get

(5.14) $$\omega(\hat{X}_2, \hat{t}_2, K_2, D(k_2)) \geq \gamma > 0$$

where $\hat{X}_2 = \bar{X}(\tilde{a}_1, \tilde{a}_2, \hat{Q}, k_2)$, $\hat{t}_2 = \bar{t}(\hat{Q})$, expressed in the $(\hat{X}, t)$ coordinate system, and $\tilde{a}_1, \tilde{a}_2$ are the constants in (1.1), (1.2), corresponding to k_2. From Lemma 5.1 observe that $\tilde{a_2}$ is a function of τ, a_1, a_2 and $\tilde{a_1} \leq \tau$. Thus $\gamma = \gamma(a_1, a_2, n)$. Since the heat equation is invariant under rotations in the space variable we deduce from (5.14) that

(5.15) $$\omega(X_1, \hat{t}_2, K_1, D(k_1)) = \omega(\hat{X}_2, \hat{t}_2, K_2, D(k_2)) \geq \gamma > 0$$

where $(X_1, \hat{t}_2)$ is the image of $(\hat{X}_2, \hat{t}_2)$ under the above transformation. From (5.10), (5.11), and the observation concerning γ we find that

(5.16) $$|X_1 - (x^*, k_1(x^*, t^*))| \leq c(a_1, a_2)\, d^{\frac{1}{2}}.$$

From (5.16) and Harnack's inequality for the heat equation, we note that (5.15) remains valid for γ small enough with $(X_1, \hat{t}_2)$ replaced by $(\tilde{X}, \tilde{t})$ where these coordinates are defined as in Lemma 4.2 with $g^* = k_1, g = f$. We also note from (5.1) that $f = k_1$ on K_1. Using these notes, and Lemmas 4.1, 4.2, with $g^* = k_1, g = f$, we find that

$$(5.17) \qquad \omega(\bar{X}, \bar{t}, F, D(f)) \geq \omega(\bar{X}, \bar{t}, K_1, D(f)) \geq \gamma_1 > 0$$

for some $\gamma_1 = \gamma_1(a_1, a_2, n)$. Since Q^*, F are arbitrary subject to $F \subset Q^* \subset R^{n-1}$ and $|F| \geq \frac{23}{24}|Q^*|$, we conclude from (5.17), Lemma 4.1, and the theorem of Coifman-Fefferman mentioned after the statement of Lemma 4.4, that Theorem 1 is valid whenever (5.8) holds and $0 < a_2 < \infty$. From (5.8) and induction we now obtain Theorem 1. □

6. Proof of Theorems 2 - 4.

In order to prove Theorems 2-4, we need to show that Lemma 4.1 remains valid when g satisfies the less restrictive hypotheses (1.4) and (1.7a). To do so we shall follow the procedure in [JK] for proving similar estimates for harmonic measure in BMO_1 domains. Let $g : R^n \to R$, and suppose $\frac{\partial g}{\partial x_j}$ exists for each j, $1 \leq j \leq n-1$, in the distributional sense. As in section 1 put

$$m_{Q^*} = m_{Q^*}(g) = |Q^*|^{-1} \int_{Q^*} g \, dx dt,$$

$$m_{j,Q^*} = m_{Q^*}\left(\frac{\partial g}{\partial x_j}\right) = \int_{Q^*} \frac{\partial g}{\partial x_j} \, dx dt,$$

$$\eta_{Q^*} = (m_{1,Q^*}, \ldots, m_{n-1,Q^*}) \, .$$

Now suppose that g satisfies (1.4) and (1.7a). We note that if $Q = Q_r(x,t) \subset Q^*$, then

$$(6.1) \qquad |\, \eta_Q - \eta_{Q^*} \,| \leq c(a_1 + a_2) \log(2 + \frac{d}{r})$$

where $c = c(n)$, as follows from standard estimates for the averages of BMO functions on rectangles. Also as (5.15) of chapter 2 we find that

$$(6.2) \qquad |g(x,t) - m_{Q^*} - \langle \eta_{Q^*}, x - x^* \rangle | \leq c(a_1 + a_2)\, d^{\frac{1}{2}}$$

for $(x,t) \in Q^*$, where $c = c(n)$ and $\langle \cdot, \cdot \rangle$ denotes the usual inner product on R^{n-1}. Given $(X,t) \in R^{n+1}$ we let $Q_r(X,t)$ be the $n+1$ dimensional rectangle defined as in (1.5) with $n-1$ replaced by n. Following Jerison and Kenig we note that (6.2) implies $D = D(g)$, $R^n \setminus \bar{D}$, satisfy the following corkscrew condition: given $(X^*, t^*) = (x^*, f(x^*, t), t^*) \in \partial D$ and $r > 0$, there exists λ, $(P_i^*, t_i^*) \in$

$D \cap Q_r(X^*, t^*)$, and $(N_i^*, \tau_i^*) \in (R^{n+1} \setminus \bar{D}) \cap \bar{Q}_r(X^*, t^*)$, $i = 1, 2$, such that

$$\text{(6.3)}\quad \begin{aligned} &(i) && \lambda^{-1} r \leq \min (t_2^* - t^*,\, t^* - t_1^*) \leq r/4, \\ &(ii) && \lambda^{-1} r \leq \min (\tau_2^* - t^*,\, t^* - \tau_1^*) \leq r/4, \\ &(iii) && (r/\lambda)^{\frac{1}{2}} \leq \min [d(\{(N_i^*, \tau_i^*)\}, \partial D), d(\{(P_i^*, t_i^*)\}, \partial D)] \,. \end{aligned}$$

Here $P_i^*, N_i^*, t_i^*, \tau_i^*$, depend on r, X^*, t^*, when $i = 1, 2$. We shall write $P_i^* = P_i^*(X^*, t^*, r)$ when we wish to indicate the dependence of P_i^* on these quantities. A similar statement holds for t_i^*, N_i^*, τ_i^*. Also in (6.3) and rest of this section, $\lambda = \lambda(a_1, a_2, n) \geq 2$ denotes a constant which depends only on a_1, a_2, n, not necessarily the same at each occurence. Next suppose $(P_i, s_i) \in D$, $i = 1, 2$, with $(s_2 - s_1)^{\frac{1}{2}} \geq \gamma^{-1} d(\{(P_1, s_1)\}, \{(P_2, s_2)\})$ for some $\gamma \geq 2$. We say that $\{ Q_{r_i}(X_i, t_i) \}_1^N$ is a <u>Harnack chain</u> from (P_1, s_1) to (P_2, s_2) with constant γ provided

(A) $(P_1, s_1) \in Q_{r_1}(X_1, t_1)$, $(P_2, s_2) \in Q_{r_N}(x_N, t_N)$, and $Q_{r_{i+1}}(X_{i+1}, t_{i+1}) \cap Q_{r_i}(X_i, t_i) \neq \emptyset$ for $i = 1, 2, \ldots, N-1$,

(B) $(n\gamma)^{-3} d(\{(X_i, t_i)\}, \partial D) \leq r_i^{\frac{1}{2}} \leq (n\gamma)^{-2} d(\{(X_i, t_i)\}, \partial D)$, when $i = 1, 2, \ldots, N$,

(C) $t_{i+1} - t_i \geq \gamma^{-4} r_i$, for $i = 1, 2, \ldots, N$.

We call N the length of this Harnack chain. Using (5.1) and (5.2) it can be shown that if $(P_i, s_i) \in D$, $i = 1, 2$, are as above, then there exists a Harnack chain from (P_1, s_1) to (P_2, s_2) with constant $\lambda = \lambda(a_1, a_2, n)$ for which

$$\text{(6.4)}\qquad N \leq \lambda \log \left(2 + \frac{\rho}{\min\{\rho_1, \rho_2\}} \right)$$

where we have put $\rho = d(\{(P_1, s_1)\}, \{(P_2, s_2)\})$, and $\rho_i = d(\{(P_i, s_i)\}, \partial D)$ for $i = 1, 2$.

A proof of (6.3), (6.4) for BMO_1 domains(using (6.1), (6.2)) can be found in [JK, p 96]. The proof is essentially unchanged for $D(g)$. To outline their proof we first observe that it suffices to prove (6.3)-(6.4) under the assumption that a_1, a_2 are small . Indeed the transformation $(x, x_n, t) \rightarrow (x, \frac{x_n}{c(a_1+a_2)}, t)$ maps $D(g)$ onto $D(g_1)$ where $g_1 = \frac{g}{c(a_1+a_2)}$ has small constants in (1.4), (1.7a), provided $c = c(n)$ is large enough. Using the inverse of this transformation it is easily checked that if (6.3) holds in $D(g_1)$ then it is also valid in $D(g)$. Moreover the inverse transformation maps a given Harnack chain satisfying (6.4) into a chain of parallelograms satisfying essentially (6.4). If (P', s') and (P'', s'') are two points of a parallelogram Λ corresponding to $Q_r(X, t)$ (a member of the Harnack chain in $D(g_1)$) and $s' - s'' \geq \lambda^{-4} r$, then there exists a Harnack chain of rectangles $\subset \Lambda$ connecting these two points with length at most $\lambda_1 = \lambda_1(a_1, a_2, n)$. From these observations we see that if (6.4) is valid in $D(g_1)$, then it is also valid in $D(g)$ for large enough λ. Thus it suffices to prove (6.3), (6.4) under the assumption that a_1, a_2 are small in (6.1), (6.2).

Now (6.2) implies for each $t \in J_n$ that the graph of $x \to g(x,t)$, $x \in J_1 \times \cdots \times J_{n-1}$ lies between two $n-1$ dimensional parallel planes at most a distance $c(a_1 + a_2)d^{\frac{1}{2}}$ apart. If a_1, a_2 are small enough it is easily seen from this geometric observation that (6.3) holds. Moreover if (P_i, s_i), $i = 1, 2$, are as in the definition of a Harnack chain and $\rho \le \min(\rho_1, \rho_2)$, then it follows from our geometric observation that (6.4) holds when a_1, a_2 are sufficiently small. Otherwise, we can use (6.3), (6.1), and our geometric observation, inductively in rectangles of the form $Q_{2^{2j}\rho_i^2}(X_i^*, t_i^*)$, $(X_i^*, t_i^*) \in \partial D$, $i = 1, 2$, where

$$d(\{(X_i^*, t_i^*)\}, \{(P_i, s_i)\}) = \rho_i \text{ for } i = 1, 2.$$

Here $2 \le j \le m_i$, for $i = 1, 2$, where m_i denotes the positive integer such that $2^{m_i} \le \rho/\rho_i < 2^{m_i+1}$. We obtain two Harnack chains in D satisfying (6.4) which connect (P_1, s_1) to (Y_1, u_1), and (Y_2, u_2) to (P_2, s_2) where

$$s_2 - s_1 \le \lambda \min\{ u_1 - s_1,\, s_2 - u_2,\, u_2 - u_1 \}$$

for some $\lambda = \lambda(a_1, a_2, n) \ge 2$. Moreover, the distance apart of $(Y_1, u_1), (Y_2, u_2)$, as well as the distance to ∂D of each point is proportional to ρ. We can then use our geometric observation again to get a Harnack chain connecting (Y_1, u_1) to (Y_2, u_2). The union of the three Harnack chains satisfies (6.4) for some $\lambda = \lambda(a_1, a_2, n) \ge 2$.

Next we state some lemmas and outline their proofs.

Lemma 6.1 *Let g satisfy (1.4), (1.7a), and suppose that u is a nonnegative solution to either the heat or adjoint heat equation in $D \cap Q_r(X^*, t^*)$ which vanishes continuously on $\partial D \cap Q_r(X^*, t^*)$. There exists $\alpha = \alpha(a_1, a_2, n), 0 < \alpha < \frac{1}{2}$, and $\lambda = \lambda(a_1, a_2, n) \ge 2$ such that*

$$u(X,t) \le \lambda \left[\frac{d(\{(X,t)\}, \{(X^*, t^*)\})^2}{r} \right]^\alpha \sup_{(Y,s) \in D \cap Q_r(X^*, t^*)} u(Y, s), \tag{6.5}$$

whenever $(X,t) \in Q_r(X^*, t^*)$.

Proof. To prove (6.5) one first uses the part of (6.3) concerning $R^{n+1} \setminus \bar{D}$ and the Poisson integral formula for a solution to the heat or adjoint heat equation in a rectangle, as in [JK, Lemma 4.1], to show that

$$\sup_{(Y,s) \in D \cap Q_{\frac{r}{\lambda}}(X^*, t^*)} u(Y, s) \le (1 - \lambda^{-1}) \sup_{(Y,s) \in D \cap Q_r(X^*, t^*)} u(Y, s)$$

for some large λ. (6.5) follows from this inequality and iteration. □

As usual we let $\tilde{\omega}(X, t, \cdot) = \tilde{\omega}(X, t, \cdot, D(g))$, be parabolic measure for $D(g)$ evaluated at $(X,t) \in D$. From Lemma 6.1 with r replaced by $\frac{r}{\lambda}$, $u = 1 - \tilde{\omega}(\cdot, \cdot, Q_{r/\lambda}(X^*, t^*) \cap \partial D)$, Harnack's inequality for the heat equation, and the Harnack chain condition (6.4), we find (see [JK, Lemma 4.2]) for some $\lambda, \lambda_2(a_1, a_2, n) \ge 2$ that

$$\tilde{\omega}(X, t, Q_{r/\lambda}(X^*, t^*) \cap \partial D) \ge \lambda_2^{-1} \tag{6.6}$$

whenever either $(X,t) \in Q_{\frac{r}{2\lambda}}(X^*, t^*) \cap D$ or $(X,t) \in D$ with $t - t^* \ge -\frac{r}{8\lambda}$ and

$$\lambda^{-1} r^{\frac{1}{2}} \le d(\{(X,t)\}, \partial D) \le d(\{(X,t)\}, \{(X^*, t^*)\}) \le \lambda r^{\frac{1}{2}}.$$

We shall also need

Lemma 6.2 *Let g, u be as in Lemma 6.1. There exists λ such that if $(X,t) \in D \cap Q_{4r/\lambda}(X^*, t^*)$, then*

$$u(X,t) \leq \lambda\, u(\hat{P}, \hat{t}) \tag{6.7}$$

where $\hat{P} = P_2^*(X^*, t^*, r)$, $\hat{t} = t_2^*(X^*, t^*, r)$ when u is a solution to the heat equation while $\hat{P} = P_1^*(X^*, t^*, r)$, $\hat{t} = t_1^*(X^*, t^*, r)$ when u is a solution to the adjoint heat equation. □

Lemma 6.2 is proved using Harnack's inequality for the heat (adjoint heat) equation, (6.4), and an argument of Caffarelli, Fabes, and Kenig as in [JK, Lemma 4.4]. We omit the details. Lemma 6.2, Harnack's inequality, and the Harnack chain condition (6.4) can also be used to prove

Lemma 6.3 *Let g satisfy (1.4), (1.7a). There exists λ such that if u is a solution to either the heat or adjoint heat equation in $D \setminus Q_{r/\lambda}(X^*, t^*)$, which vanishes continuously on $\partial D \setminus \bar{Q}_{r/\lambda}(X^*, t^*)$, then*

$$u(X,t) \leq \lambda\, u(\hat{P}, \hat{t}) \tag{6.8}$$

whenever $(X,t) \in D \setminus Q_{\frac{3r}{2\lambda}}(X^*, t^*)$. Here, $(\hat{P}, \hat{t})$ is as in Lemma 6.2. □

To prove Lemma 6.3 we note from the maximum principle for the heat equation, Harnack's inequality, and the Harnack chain condition (6.4), that it suffices to prove Lemma 6.3 for points on $D \cap \partial Q_{\frac{3r}{2\lambda}}(X^*, t^*)$ which are near ∂D. However, these points are covered by Lemma 6.2.

Let $G(\cdot, \cdot, X_0, t_0)$ be Green's function in D for the heat equation with pole at $(X_0, t_0) \in D$. We note that $G(X_0, t_0, \cdot, \cdot)$ is Green's function in D for the adjoint heat equation with pole at (X_0, t_0) (see [W, section 2] for definitions and properties of G). With this notation we are ready to state the next lemma.

Lemma 6.4 *Let g satisfy (1.4) and (1.7a). Suppose that $(X,t) \in D$ with $t - t^* \geq r$. There exists λ such that*

$$\tilde{\omega}(X, t, \partial D \cap Q_{r/\lambda}(X^*, t^*)) \leq \lambda\, r^{\frac{n}{2}}\, G(X, t, P_1^*, t_1^*) \tag{6.9}$$

where (P_1^*, t_1^*) is defined as in (6.3) relative to X^*, t^*, r. □

(6.9) follows from standard p.d.e estimates using the equality

(6.10)
$$-\int_{\partial D} \psi\, d\tilde{\omega}(X, t, \cdot, \cdot) = \int_D [\langle \nabla_Y G(X, t, \cdot, \cdot), \nabla_Y \psi \rangle + \frac{\partial \psi}{\partial t} G(X, t, \cdot, \cdot)]\, dY ds$$

where $\psi \in C_0^\infty(R^{n+1})$ vanishes in a neighborhood of (X,t). This equality is derived by first using Sard's theorem and the definition of parabolic measure to get similar equalities in $\{(Y,s) : G(X,t,Y,s) > \delta\}$. Second, letting $\delta \to 0$ through a certain sequence, taking a weak limit, and using the definition of $\tilde{\omega}$, we obtain (6.10).

We refer the reader to [FGS, Lemma 1] for a complete proof of Lemma 6.4 using similar p.d.e estimates.

The following lemma can also be found in [FGS, Theorem 3] for Lip $(1,\frac{1}{2})$ domains.

Lemma 6.5 *Let g satisfy (1.4) and (1.7a). There exists λ such that if u is a nonnegative solution and v a positive solution to either the heat or adjoint heat equation in $Q_r(X^*,t^*)\cap D$, which vanish continuously on $\partial D\cap Q_r(X^*,t^*)$, then*

$$\frac{u(Y,s)}{v(Y,s)}\le\lambda\frac{u(\bar P,\bar t)}{v(\tilde P,\tilde t)} \tag{6.11}$$

whenever $(Y,s)\in Q_{r/\lambda^3}(X^*,t^*)$. Here $\bar P=P_2^*$, $\bar t=t_2^*$, $\tilde P=P_1^*$, and $\tilde t=t_1^*$ when u,v are solutions to the heat equation while $\bar P=P_1^*$, $\bar t=t_1^*$, $\tilde P=P_2^*$, and $\tilde t=t_2^*$ when u,v are solutions to the adjoint heat equation.

Proof. We outline the proof of Fabes, Garofalo, and Salsa, when u,v are solutions to the heat equation in $Q_r(X^*,t^*)\cap D$. First we use a well known covering argument to obtain $\{Q_{\frac{r}{\lambda^3}}(Y_i,s_i)\}$ with

(a) $\quad (Y_i,s_i)\in\partial D\cap\{(X,t)\in\partial Q_{\frac{r}{\lambda^2}}(X^*,t^*):t^*-t=\frac{r}{2\lambda^2}\}$ for each i,

(b) $\quad Q_{\frac{r}{10^n\lambda^3}}(Y_i,s_i)\cap Q_{\frac{r}{10^n\lambda^3}}(Y_j,s_j)=\emptyset$ for $i\neq j$,

(c) $\quad \partial D\cap\{(X,t)\in\partial Q_{\frac{r}{\lambda^2}}(X^*,t^*):t^*-t=\frac{r}{2\lambda^2}\}\subset\cup Q_{\frac{r}{\lambda^3}}(Y_i,s_i)$.

Using (6.6), the maximum principle for the heat equation, Harnack's inequality, (6.4), and Lemma 6.2 we deduce that for $(Y,s)\in Q_{\frac{r}{\lambda^2}}(X^*,t^*)\cap D$

$$u(Y,s)\le\lambda\,u(P_2^*,t_2^*)\sum_i\tilde\omega(Y,s,Q_{\frac{r}{\lambda^3}}(Y_i,s_i)\cap\partial D)\,. \tag{6.12}$$

From (b) we see that the number of rectangles in the above sequence is at most $c\lambda^{\frac{n+2}{2}}$. Using this fact, Lemma 6.4 with $r/\lambda^2,(Y_i,s_i)$, replacing $r,(X^*,t^*)$, respectively; (6.12), (6.4), and Harnack's inequality for the adjoint heat equation, we find for $(Y,s)\in Q_{\frac{r}{\lambda^3}}(X^*,t^*)\cap D$ that

$$u(Y,s)\le\lambda\,r^{\frac{n}{2}}\,u(P_2^*,t_2^*)\,G(Y,s,P_1^*,t_1^*). \tag{6.13}$$

We note that if W denotes the fundamental solution to the heat equation (i.e the Green's function for R^{n+1}) with pole at (P_1^*,t_1^*), then $G(\cdot,\cdot,P_1^*,t_1^*)\le W$ and for large λ

$$W-(\lambda/r)^{\frac{n}{2}}\le G(\cdot,\cdot,P_1^*,t_1^*)\ \text{ in } E(\lambda)=\{(x,t):W(x,t)>(\lambda/r)^{\frac{n}{2}}\}.$$

Hence if $\Omega=\{(Y,s):G(Y,s,P_1^*,t_1^*)>(\lambda^2/r)^{\frac{n}{2}}\}$, then

$$E(\lambda^3)\subset\Omega\subset E(\lambda^2) \tag{6.14}$$

for large λ and

$$2d(P_1^*,t_1^*)\ge d(\Omega,\partial D)\ge r^{\frac{1}{2}}/\lambda^{\frac{5}{4}}\,. \tag{6.15}$$

Using (6.15), Harnack's inequality, and the maximum principle for the heat equation once again, we conclude that

$$\lambda v(Y,s) \geq v(P_1^*,t_1^*)\,(r/\lambda^2)^{\frac{n}{2}}\, G(Y,s,P_1^*,t_1^*) \tag{6.16}$$

when $(Y,s) \in Q_{\frac{r}{\lambda^3}}(X^*,t^*) \cap D$. From (6.16) and (6.13) we see that Lemma 6.5 is true. □

Finally we state an analogue of Lemma 4.1 for our domains.

Lemma 6.6 *Let g satisfy (1.4) and (1.7a). Suppose that $(X,t) \in D$ with $|X-X^*|^2 \leq A(t-t^*)$ for some fixed $A \geq 2$. There exists λ, $\mu = \mu(a_1,a_2,A,n) \geq 2$ such that if $\frac{r}{\lambda} \leq t - t^*$, then*

$$\tilde{\omega}(X,t,\partial D \cap Q_{\frac{r}{\lambda}}(X^*,t^*)) \leq \mu\tilde{\omega}(X,t,\partial D \cap Q_{\frac{r}{2\lambda}}(X^*,t^*)) \tag{6.17}$$

Moreover, if E is a Borel subset of $\partial D \cap Q_{\frac{r}{\lambda}}(X^*,t^*)$, then

$$\mu^{-1}\,\tilde{\omega}(P_2^*,t_2^*,E) \leq \frac{\tilde{\omega}(X,t,E)}{\tilde{\omega}(X,t,Q_{\frac{r}{\lambda}}(X^*,t^*) \cap \partial D)} \leq \mu\tilde{\omega}(P_2^*,t_2^*,E) \tag{6.18}$$

where P_2^*, t_2^* are as in (6.3) relative to r, X^*, t^*.

Proof. We note that (6.17) follows from (6.6) and (6.18) with $E = Q_{\frac{r}{2\lambda}}(X^*,t^*)$. Thus we shall only sketch the proof of (6.18). To do this we follow [B2, Lemma 2.2] and first assume that (X,t) lies in a rectangle $Q_{\frac{\tau}{\lambda^3}}(\hat{X},\hat{t})$ with

(6.19)

(+) $\quad \tau \geq \frac{r}{100\lambda^2}$ and $(\hat{X},\hat{t}) \in \partial D$,

(++) $\quad Q_\tau(\hat{X},\hat{t}) \cap Q_{r/\lambda}(X^*,t^*) = \emptyset$,

(+++) $\quad \max(\frac{\tau}{\lambda}, \frac{r}{\lambda}) \leq \hat{t} - t^* \leq d^2\left(\{(\hat{X},\hat{t})\},\{(X^*,t^*)\}\right) \leq \lambda\tau.$

Then from Lemma 6.5 with r, X^*, t^* replaced by $\tau, \hat{X}, \hat{t}$, *respectively*; $u = \tilde{\omega}(\cdot,\cdot,E)$, $v = \tilde{\omega}(\cdot,\cdot,Q_{\frac{r}{\lambda}}(X^*,t^*))$, we deduce

$$\frac{\tilde{\omega}(X,t,E)}{\tilde{\omega}(X,t,Q_{\frac{r}{\lambda}}(X^*,t^*))} \leq \lambda\frac{\tilde{\omega}(P_2,t_2,E)}{\tilde{\omega}(P_1,t_1,Q_{\frac{r}{\lambda}}(X^*,t^*))} \tag{6.20}$$

where $P_1 = P_1^*(\hat{X},\hat{t},\tau), t_1 = t_1^*(\hat{X},\hat{t},\tau)$ and P_2, t_2 have a similar meaning. Also from (6.8) of Lemma 6.3, Harnack's inequality, (6.4), and the maximum principle for the heat equation we deduce that

$$\lambda^{-1}\tilde{\omega}(P_2,t_2,F) \leq \tilde{\omega}(P_1,t_1,F) \leq \lambda\tilde{\omega}(P_2,t_2,F) \tag{6.21}$$

whenever $F \subset Q_{\frac{r}{\lambda}}(X^*,t^*)$ is Borel. Applying deduction (6.21) with $F = E$, $Q_{\frac{r}{\lambda}}(X^*,t^*)$, respectively we find in view of (6.20) that

$$\frac{\tilde{\omega}(X,t,E)}{\tilde{\omega}(X,t,Q_{\frac{r}{\lambda}}(X^*,t^*))} \leq \lambda\frac{\tilde{\omega}(P_2,t_2,E)}{\tilde{\omega}(P_2,t_2,Q_{\frac{r}{\lambda}}(X^*,t^*))}. \tag{6.22}$$

Hence in the case we are considering it suffices to prove the righthand inequality in Lemma 6.6 for (P_2, t_2). To do this we first note that if $\frac{2r}{\lambda} \le \hat{t} - t^* \le r$, then from (+++), Lemma 6.3, (6.4), and Harnack's inequality it follows that $\tilde{\omega}(P_2, t_2, F)$ and $\tilde{\omega}(P_2^*, t_2^*, F)$ are comparable whenever $F \subset Q_{\frac{r}{\lambda}}(X^*, t^*) \cap \partial D$ is Borel. Here P_2^*, t_2^*, are defined relative to (X^*, t^*), r, as in (6.3). Thus we assume that $\hat{t} - t^* > r$. Using an argument similar to the proof of (6.13), we find for $u = \tilde{\omega}(\cdot, \cdot, E)$, that

$$\text{(6.23)} \qquad \lambda\tilde{\omega}(P_2, t_2, E) \le r^{\frac{n}{2}}\,\tilde{\omega}(P_2^*, t_2^*, E)\, G(P_2, t_2, P_1^*, t_1^*).$$

Also an argument as in (6.16) gives for $v = \tilde{\omega}(\cdot, \cdot, Q_{\frac{r}{\lambda}}(X^*, t^*))$ that

$$\text{(6.24)} \quad \lambda\tilde{\omega}(P_2, t_2, Q_{\frac{r}{\lambda}}(X^*, t^*)) \ge r^{\frac{n}{2}}\,\tilde{\omega}(P_2^*, t_2^*, Q_{\frac{r}{\lambda}}(X^*, t^*))\, G(P_2, t_2, P_2^*, t_2^*).$$

We claim that

$$\text{(6.25)} \qquad \lambda^{-1}\, G(P_2, t_2, P_2^*, t_2^*) \le G(P_2, t_2, P_1^*, t_1^*) \le \lambda\, G(P_2, t_2, P_2^*, t_2^*)\,.$$

The lefthand inequaliy in (6.25) is a consequence of (6.4) and Harnack's inequality for the adjoint heat equation. To prove the righthand inequality in this claim, we note from (1.4) that if $h = t_2^* - t_1^* + \frac{r}{\lambda^2}$, then the transformation

$$\Phi(Y, s) = (y, y_n + ca_2\, h^{\frac{1}{2}}, s + h)$$

maps D into D for $c = c(n)$ large enough. Also if Ω is as in (6.15) and $(\tilde{P}_2, \tilde{t}_2) = \Phi(P_1^*, t_1^*)$, then as in (6.14) we find $\Phi(\Omega) \subset \Omega_1$ for large λ, where

$$\Omega_1 = \{(Y, s) : G(Y, s, \tilde{P}_2, \tilde{t}_2) > (\lambda/r)^{\frac{n}{2}}\}.$$

Using these facts, as well as the maximum principle for the heat equation and its invariance under translations, we get

$$\text{(6.26)} \qquad G(Y, s, P_1^*, t_1^*) \le c\,\lambda\, G(\Phi(Y, s), \tilde{P}_2, \tilde{t}_2)$$

whenever $(Y, s) \in D \setminus \Omega$. Also it follows from Lemma 6.3 as in (6.21), Harnack's inequality for the adjoint heat equation and (6.4) that

$$G(\Phi(P_2, t_2), \tilde{P}_2, \tilde{t}_2) \le \lambda\, G(P_2, t_2, \tilde{P}_2, \tilde{t}_2) \le \lambda^2\, G(P_2, t_2, P_2^*, t_2^*).$$

Combining this inequality with (6.26) evaluated at (P_2, t_2) we obtain claim (6.25). From (6.23)-(6.25) and (6.6), we deduce

$$\frac{\tilde{\omega}(P_2, t_2, E)}{\tilde{\omega}(P_2, t_2, Q_{\frac{r}{\lambda}}(X^*, t^*) \cap \partial D)} \le \lambda\,\tilde{\omega}(P_2^*, t_2^*, E)\,.$$

From the above inequality and (6.22) we see that the righthand inequality in (6.18) is valid when case (6.19) occurs.

If (6.19) is false, then from the assumptions on (X, t) in Lemma 6.6 and a use of Lemma 6.3 as in (6.21), we see there exists (X_1, t_1) such that

$$\text{(6.27)} \qquad \mu^{-1}\,\tilde{\omega}(X_1, t_1, F) \le \tilde{\omega}(X, t, F) \le \mu\,\tilde{\omega}(X_1, t_1, F)$$

whenever $F \subset Q_{\frac{r}{\lambda}}(X^*, t^*)$. Also there exist corresponding $(\hat{X}_1, \hat{t}_1)$, τ_1, such that (X_1, t_1) lies in $Q_{\tau_1/\lambda^3}(\hat{X}_1, \hat{t}_1)$ and (6.19) holds with $(\hat{X}, \hat{t})$ replaced by $(\hat{X}_1, \hat{t}_1)$. We can apply the previous argument to (X_1, t_1) to get the righthand inequality in (6.18) with (X,t) replaced by (X_1,t_1). From (6.27) we conclude that the righthand inequality in (6.18) is true. The lefthand inequality is proved similarly. This concludes the proof of Lemma 6.6 . $\square$

Armed with Lemma 6.6, we could proceed as in [B2] and prove a version of Lemma 4.2 for our domains. However, the argument is somewhat simpler if we proceed as in [JK, section 10] and as in Lemma 5.1 of chapter 2. For this purpose let $f: R^n \to R$ satisfy (1.4), (1.7) and put $D = D(f)$. We note that (1.4), (1.7) imply

$$\int_{J_n}\left(\int_{Q^*} \frac{(f(x,s)-f(x,t))^2}{(s-t)^2}\,dxdt\right) ds \le c(a_1+a_2)^2\,|Q^*| \tag{6.28}$$

whenever $d > 0$ and $(x^*,t^*) \in R^n$. The proof of (6.28) is similar to the proof of (2.17) (see [S]). Recall from section 1 that

$$\rho(z,t) = (z, f(z,t), t),\ (z,t) \in R^n\ ,$$

$$\sigma(F) = \int_{\rho^{-1}(F)} \sqrt{1+|\nabla_x f|^2}(x,t)\,dxdt,$$

whenever $F \subset \partial D$ is Borel. Let $r > 0$ and $(X^*,t^*) \in \partial D$, where $X^* = (x^*, f(x^*,t^*))$. Then in (5.8) of chapter 2 we proved that

$$\lambda^{-1} r^{\frac{n+1}{2}} \le \sigma(Q_r(X^*,t^*) \cap \partial D) \le \lambda r^{\frac{n+1}{2}}\,. \tag{6.29}$$

Writing $\rho^{-1}(Q_r(X^*,t^*) \cap \partial D)$ as a union of rectangles, using (6.28) and weak type estimates as in the argument leading up to (5.24) in chapter 2, we find the existence of a closed set $E \subset \rho^{-1}(Q_r(X^*,t^*) \cap \partial D)$ with $r^{\frac{n+1}{2}} \le \lambda\,\sigma(E)$. Moreover if $K = \{t : |t-t^*| < r\}$, then for $(x,t) \in E$

$$\int_K \frac{(f(x,t)-f(x,s))^2}{(s-t)^2}\,ds \le \lambda \log(2+|\eta_{\tilde Q}|)\,,$$

where $\tilde Q = Q_{8r}(x^*,t^*)$. Next set

$$H(x,t) = f(x,t) - m_{\tilde Q} - \langle \eta_{\tilde Q}, x - x^*\rangle$$

when $(x,t) \in Q_{16r}(x^*,t^*)$ and $H = 0$, otherwise in R^n . Let

$$M(|\nabla_x H|)(x,t) = \sup_Q \left[|Q|^{-1} \int_Q |\nabla_y H|(y,s)dyds \right]$$

where the supremum is over all $Q = Q_d(x,t)$ whenever $(x,t) \in R^n$. Note that $M(|\nabla_x H|)$ is the Hardy Littlewood maximal function of $|\nabla_x H|$ on rectangles. We shall need the following lemma.

Lemma 6.7 *If* $\lambda_3 = \lambda_3(a_1, a_2, n)$ *is large enough, then there exists a closed set* E_1 *and* $\phi: R^n \to R$ *satisfying (1.1) - (1.4), with* a_1, a_2 *replaced by* $\lambda_3 \log(2 + |\eta_{\tilde{Q}}|)$. *Also,*

(6.30)

$$(A)\quad E_1 \subset E \cap \{(x,t) : M(|\nabla_x H|)(x,t) \le \lambda_3 \log(2 + |\eta_{\tilde{Q}}|)\},$$

$$(B)\quad \lambda_3\, \sigma(\rho(E_1)) \ge r^{\frac{n+1}{2}},$$

$$(C)\quad \phi = H \quad on \quad E_1,$$

$$(D)\quad \int_K \frac{(\phi(x,t) - \phi(x,s))^2}{(s-t)^2}\, ds \le \lambda_3 (1 + |\eta_{\tilde{Q}}|)\, [\log(2 + |\eta_{\tilde{Q}}|)]^2, (x,t) \in E_1,$$

$$(E)\quad \text{If } (x,t) \in \tilde{Q}, \text{ then } |\phi(x,t) - H(x,t)| \le \lambda \log(2 + |\eta_{\tilde{Q}}|)\, d(\{(x,t)\}, E_1). \square$$

Lemma 6.7 is just Lemma 5.1 in chapter 2 except for (E) which is implied by the proof of this lemma (see (5.31) of chapter 2). We note that the proof of Lemma 6.7 is similar to the proof of Lemma 3.1 in this chapter.

Proof. Using Lemmas 6.6 and 6.7, we can now prove Theorem 2. Given $r > 0$ and $(X^*, t^*) \in \partial D$, let $\Lambda \subset \bar{Q}_r(X^*, t^*)$ be Borel and suppose that

$$\sigma(\Lambda) \ge (1 - \lambda_4^{-1})\sigma(Q_r(X^*, t^*)). \tag{6.31}$$

If $\lambda_4 = \lambda_4(a_1, a_2, n)$ is large enough we claim there exists $\lambda_5 = \lambda_5(a_1, a_2, n)$ such that

$$\tilde{\omega}(P_2, t_2, \Lambda) \ge \lambda_5^{-1}. \tag{6.32}$$

Here, $P_2 = P_2^*(X^*, t^*, \lambda r)$, $t_2 = t_2^*(X^*, t^*, \lambda r)$, and λ is as in Lemma 6.6. Theorem 2 follows from (6.32), Lemma 6.6 with r replaced by λr, and a mild generalization of the theorem of Coifman-Fefferman [CF] mentioned in section 5 (see [JK], section 10 for references).

To prove our claim let E_1 be as in Lemma 6.7 and set $\delta(x,t) = d(\{(x,t)\}, E_1)$ when $(x,t) \in R^n$. Also put

$$\psi(x,t) = \phi(x,t) + \lambda_6 \log(2 + |\eta_{\tilde{Q}}|)\, \delta(x,t),\ (x,t) \in R^n .$$

We note that δ satisfies (1.1), (1.4) with constant $c = c(n)$, and this function is also Lipschitz in t on $R^n \setminus E_1$, with

$$\left|\frac{\partial \delta}{\partial t}\right|(x,t) \le c\, \delta(x,t)^{-\frac{1}{2}}, \text{ for a.e } (x,t) \in R^n \setminus E_1.$$

Using these facts it follows as in the proof of Lemma 3.1 that δ also satisfies (1.2), (1.3) with comparable constants to those in (1.1), (1.4). We note that essentially $h_1 = \delta$ is the extension of $h = 0$ in Lemma 3.1 .

Next we let

$$\Psi(x,t) = \psi(x,t) + m_{\tilde{Q},f} + \langle \eta_{\tilde{Q},f}, x - x^* \rangle,\ (x,t) \in R^n ,$$

and note from (6.2) with $g = f$, Lemma 6.7, and our choice of ψ that for λ_6 large enough

(6.33) $$\Psi(x,t) \geq f(x,t),\ (x,t) \in \tilde{Q}.$$

From a Phragmen-Lindelöf type maximum principle for bounded solutions to the heat equation, 6.30 (C), and (6.33) we obtain that

(6.34) $$\tilde{\omega}(X,t,\Lambda \cap \rho(E_1), D(\Psi)) \leq \tilde{\omega}(X,t,\Lambda \cap \rho(E_1), D(f))$$

whenever $(X,t) \in D(\Psi)$ with $t - t^* < 4r$. We now consider two cases. First suppose that

(6.35) $$|\eta_{\tilde{Q}}| \leq \lambda_0$$

for some fixed λ_0 to be determined later. In this case we see from Lemma 6.7 that Ψ satisfies (1.1)-(1.4) with constant λ. Thus we can use Theorem 1 in $D(\Psi)$. From Theorem 1, (6.30)(B), and (6.29) we see for λ_4 large enough in (6.31) (depending on λ_0) that there exists $(\tilde{P},\tilde{t}) \in D(\Psi)$ with $\tilde{t} - t^* = 2r$, $\frac{r}{\lambda} \leq d(\{(\tilde{P},\tilde{t})\}, \partial D(\Psi)) \leq r\lambda$, and

$$\tilde{\omega}(\tilde{P},\tilde{t},\Lambda \cap \rho(E_1), D(\Psi)) \geq \lambda^{-1}.$$

From (6.34) we see that this inequality remains valid with $D(\Psi)$ replaced by $D(f)$. Using this fact, Lemma 6.3, (6.4), and Harnack's inequality we deduce that (6.32) is true .

On the other hand if (6.35) is false, we proceed as in chapter 2 from (5.36) to the end of section 5. Choose a rotation U of R^{n-1} in such a way that if $x = Ux'$, then $\eta_{\tilde{Q}} = |\eta_{\tilde{Q}}|\, U e_1$, where $e_1 = (1, ..., 0)$ in the rotated coordinate system . Put $\Psi_1(x',t) = \Psi(Ux',t)$ and note from Lemma 6.7, as well as our remarks about δ, that

(6.36)
$$(i) \qquad \left|\frac{\partial \Psi_1}{\partial x_1'} - |\eta_{\tilde{Q}}|\right| \leq \lambda \log(2 + |\eta_{\tilde{Q}}|),$$
$$(ii) \qquad \left|\frac{\partial \Psi_1}{\partial x_j'}\right| \leq \lambda \log(2 + |\eta_{\tilde{Q}}|),\ 2 \leq j \leq n-1.$$

We now proceed as in Lemma 5.1 of section 5. Let $\theta = -\arctan(|\eta_{\tilde{Q}}|)$ and set

(6.37)
$$\begin{aligned} x_1' &= \cos\theta\, \hat{x}_1 + \sin\theta\, \Psi_2(\hat{x}_1, ..., \hat{x}_{n-1}, t), \\ x_j' &= \hat{x}_j, 2 \leq j \leq n-1, \\ \Psi_1(x_1', ..., x_{n-1}', t) &= -\sin\theta\, \hat{x}_1 + \cos\theta\, \Psi_2(\hat{x}_1, ..., \hat{x}_{n-1}, t). \end{aligned}$$

As in (5.3)-(5.6) with $\phi = \theta$ it follows that if λ_0 is large enough, then Ψ_2 exists satisfying (6.37) and

(6.38) $$|\nabla_{\hat{x}} \Psi_2|(\hat{x},t) \leq \frac{\lambda \log(2 + |\eta_{\tilde{Q}}|)}{1 + |\eta_{\tilde{Q}}|},$$

whenever $(\hat{x},t) \in R^n$. Fix λ_0 so that Ψ_2 exists. Using (6.37) - (6.38) and Lemma 6.7, it also follows that

(6.39) $$|\Psi_2(\hat{x},t) - \Psi_2(\hat{x},s)| \leq \frac{\lambda[\log(1 + |\eta_{\tilde{Q}}|)]^2}{1 + |\eta_{\tilde{Q}}|} |s-t|^{\frac{1}{2}},\ (\hat{x},t), (\hat{x},s) \in R^n .$$

Moreover if $\hat{E}_1$ denotes the image of E_1 under the transformation $(x,t)\to(x',t)\to(\hat{x},t)$, then from our choice of θ and (6.30)(D), we deduce that

$$\int_K \frac{(\Psi_2(\hat{x},t)-\Psi_2(\hat{x},s))^2}{(s-t)^2}\,ds \le \lambda\frac{[\log(2+|\eta_{\hat{Q}}|)]^2}{1+|\eta_{\hat{Q}}|^2}\int_K \frac{(\psi(x,t)-\psi(x,s))^2}{(s-t)^2}\,ds \tag{6.40}$$

$$\le \lambda\frac{[\log(2+|\eta_{\hat{Q}}|)]^4}{1+|\eta_{\hat{Q}}|}$$

when $(\hat{x},t)\in\hat{E}_1$. From (6.38) - (6.40) we see that Lemma 3.1 can be used with $\Psi_2 = h$, $E=\hat{E}_1$. Applying this lemma we get Ψ_3 satisfying (1.1) - (1.4) with a_1, a_2, replaced by $\frac{\lambda\,[\log(2+|\eta_{\hat{Q}}|)]^2}{1+|\eta_{\hat{Q}}|^{\frac{1}{2}}}$ and $\Psi_2 = \Psi_3$ on $\tilde{E}$.

Next given $(\hat{X},t)\in R^{n+1}$, define (X',t) by

$$x_1' = \cos\theta\,\hat{x}_1 + \sin\theta\,\hat{x}_n,$$
$$x_j' = \hat{x}_j,\ 2\le j\le n-1,$$
$$x_n' = -\sin\theta\,\hat{x}_1 + \cos\theta\,\hat{x}_n\,.$$

Also define (X,t) corresponding to (X',t) by $x=Ux'$, $x_n = x_n'$. Clearly the transformation $(\hat{X},t)\to(X',t)\to(X,t)$ is just a rotation in the space variable. Moreover from (6.37) we see that this transformation maps $D(\Psi_2)$ expressed in the $(\hat{X},t)$ coordinate system onto $D(\Psi)$ expressed in the (X,t) coordinate system. Let $\tilde{\rho}(\hat{x},t) = (\hat{x},\Psi_3(\hat{x},t),t)$ and put

$$\tilde{\sigma}(F) = \int_{\tilde{\rho}^{-1}(F)} \sqrt{1+|\nabla_{\hat{x}}\Psi_3|^2}(\hat{x},t)\,d\hat{x}dt$$

when F is a Borel set contained in $\partial D(\Psi_3)$. Let $\hat{\Lambda}$ be the image of Λ under the transformation $(X,t)\to(X',t)\to(\hat{X},t)$. From our construction we have

$$\sigma(\Lambda\cap\rho(E_1)) = \tilde{\sigma}(\tilde{\rho}(\hat{E}_1)\cap\hat{\Lambda}) \le \lambda\,|\,\tilde{\rho}^{-1}(\hat{\Lambda})\cap\hat{E}_1)|. \tag{6.41}$$

Applying Theorem 1 in $D(\Psi_3)$, using (6.41), (6.30) (B), and (6.29), we see that if λ_4 is large enough in (6.31), then there exists λ_7, $\lambda_8\ge 2$, as well as $(\hat{P},\hat{t})\in D(\Psi_3)$ with $\hat{t}-t^* = 2r$ and

$$\lambda_7\, r^{\frac{1}{2}} \le d\left(\{\hat{P},\hat{t}\},\partial D(\Psi_3)\right) \le \lambda_7^2\, r^{\frac{1}{2}}$$

such that

$$\tilde{\omega}(\hat{P},\hat{t},\hat{\Lambda}\cap\tilde{\rho}(\hat{E}_1),)\,D(\Psi_3)) \ge \lambda_8^{-1}\,. \tag{6.42}$$

From Lemma 4.2 with $g=\Psi_2$, $g^*=\Psi_3$, Lemma 4.1, and (6.38)-(6.39), we see that if λ_7, λ_8 are large enough then (6.42) holds with $D(\Psi_3)$ replaced by $D(\Psi_2)$. Moreover since the heat equation is invariant under rotations in the space variable we have

$$\tilde{\omega}(P,\hat{t},\Lambda\cap\rho(E_1),D(\Psi)) = \tilde{\omega}(\hat{P},\hat{t},\hat{\Lambda}\cap\tilde{\rho}(\hat{E}_1),D(\Psi_2)) \ge \lambda_8^{-1} \tag{6.43}$$

where $(P,\hat{t})$ is the image of $(\hat{P},\hat{t})$ under the transformation $(\hat{X},\hat{t})\to(X',\hat{t})\to(X,\hat{t})$. From (6.43) and (6.34) we see that

$$\tilde{\omega}(P,\hat{t},\Lambda\cap\rho(E_1),D(f)) \ge \lambda_8^{-1}. \tag{6.44}$$

From (6.44) and the backward Harnack argument used in the first case we find that (6.32) is valid when (6.35) is false. Hence claim (6.32) is valid in either case. Theorem 2 now follows from the remarks after this claim. □

Proof. Next we prove Theorem 3. Let f be as in the statement of this theorem. We note that (6.2), (6.29) are still valid for f since (1.4), (1.7a), were only used in the proof of these inequalities. To prove Theorem 3 it clearly suffices to show that if $F \subset Q^*$ is Borel, then

$$\sigma(\rho(F)) > 0 \rightarrow \tilde{\omega}(X, t, \rho(F), D(f)) > 0. \tag{6.45}$$

To prove (6.45) we define $H, \tilde{Q}$, as in the display preceding (6.30) with r replaced by d. Observe from the hypotheses of Theorem 3 and weak type estimates that there exists a Borel set $E \subset F$ such that $\sigma(E) \geq \frac{1}{2}\sigma(F)$ and

$$\int_{J_n} \frac{(f(x,t) - f(x,s))^2}{(s-t)^2}\, ds \leq k_1 \log(2 + |\eta_{\tilde{Q}}|) < \infty \tag{6.46}$$

for some large k_1, whenever $(x, t) \in E$. Using the above note, (6.46), and arguing as in Lemma 5.1 of chapter 2, we see that Lemma 6.7 remains valid with λ_3 replaced by k_2 where k_2 depends on $k_1, \sigma(F), d, a_1, a_2$ and n. With ϕ so constructed define ψ, Ψ, as in the displays preceding (6.33) with λ_6 replaced by k. If k is large enough we find from Lemma 6.7, as previously, that (6.33) and (6.34) are valid. Since Ψ satisfies the hypotheses of Theorem 1, we can conclude from this theorem for suitably large A that

$$\tilde{\omega}(X, t, \rho(E_1), D(\Psi)) > 0 \tag{6.47}$$

whenever $(X, t) \in D(\Psi)$. From (6.47), (6.34) we see that

$$\tilde{\omega}(X_1, t_1, \rho(E_1), D(f)) > 0 \tag{6.48}$$

whenever $(X_1, t_1) \in D(f)$ and $t_1 - t^* \leq 4d$. Now it follows easily from Harnack's inequality and the maximum principle for bounded solutions to the heat equation that if

$$\tilde{\omega}(X_2, t_2, \rho(E_1), D(f)) = 0$$

for any $(X_2, t_2) \in D(f)$ with $t_2 - t^* > d/2$, then

$$\tilde{\omega}(X, t, \rho(E_1), D(f)) \equiv 0$$

whenver $(X, t) \in D(f)$ with $t - t^* > d/2$. From this fact and (6.48) we conclude first that (6.45) holds and thereupon that Theorem 3 is true. □

Proof. To prove Theorem 4 it clearly suffices to show that if $E \subset Q^*$ is a Borel set satisfying (6.46) and

$$M(|\nabla_x f|)(x, t) \leq k_1,\ (x, t) \in E,$$

then

$$\tilde{\omega}(X, t, \rho(E), D(f)) > 0 \rightarrow \sigma(E) > 0 \tag{6.49}$$

To this end we again note from the proof of Lemma 5.1 in chapter 2 that Lemma 6.7 holds with λ_3 replaced by k_2 and $r = d$. Moreover for k_2 suitably large we can arrange it so that

$$\tilde{\omega}(X, t, \rho(E_1), D(f)) > 0. \tag{6.50}$$

If ϕ is as in this lemma put

$$\hat{\psi}(x,t) = \phi(x,t) - k \log(2 + |\eta_{\tilde{Q}}|)\, \delta(x,t),$$

$$\hat{\Psi}(x,t) = \hat{\psi}(x,t) + m_{\tilde{Q},f} + \langle \eta_{\tilde{Q},f}, x - x^* \rangle,$$

when $(x,t) \in R^n$. If k is large enough it follows from Lemma 6.7(as previously) that $\hat{\Psi}(x,t) \leq f(x,t)$ when $t - t^* \leq 4d$. From this inequality we deduce as in (6.34) that

$$\tilde{\omega}(X_1, t_1, \rho(E_1), D(\hat{\Psi})) \geq \tilde{\omega}(X_1, t_1, \rho(E_1), D(f)) \tag{6.51}$$

whenever $(X_1, t_1) \in D(f)$ and $t_1 - t^* \leq 4d$. From (6.50), (6.51), and the statement following (6.48) we deduce that

$$\tilde{\omega}(X_1, t_1, \rho(E_1), D(\hat{\Psi})) > 0$$

if $t_1 - t^* = d$ and $(X_1, t_1) \in D(f)$. Finally since $\hat{\Psi}$ satisfies the hypotheses of Theorem 2 we can apply this theorem to conclude in view of the above inequality that

$$\sigma(E) \geq \sigma(E_1) > 0.$$

The proof of Theorem 4 is now complete. □

7. Closing Remarks.

This chapter has been mainly concerned with the mutual absolute continuity of parabolic measure and a certain projective Lebesgue measure. Therefore since Theorem 1 for graph domains also holds locally (thanks to Lemma 4.2), we did not bother to define the analogue of Lipschitz or C^1 domains for the heat equation. As regards the Dirichlet and Neumann problems, we mention one very interesting problem for further research . Namely, can the small norm condition in Theorems 5 and 6 be improved upon if we only ask for the solvability of the Dirichlet and Neumann problems in $L^2(R^n)$? For example, perhaps the assumption a_1 small is unnecessary in Theorems 5 and 6. As evidence for this query, we note that Brown used a Rellich type inequality in [B1] to show that Theorems 5 and 6 are valid on Lipschitz cylinders. Also, the first author believes that a close examination of the operators in chapter 1 along with Brown's technique can be used to show that Theorems 5 and 6 remain valid for $L^2(R^n)$ when f has the form $f(x,t) = B_1(t)\phi(x) + B_2(t)$ where ϕ is Lipschitz on R^{n-1} and B_i satisfies

$$\sup_{\{|s-t|\leq\tau\}} |B_i(s) - B_i(t)| \leq \gamma(\tau)$$

for $i = 1, 2$, and $0 < \tau < \infty$. Here, $\int_0^\infty \gamma^2(\tau)\,\tau^{-2}\,d\tau < \infty$.

From the above discussion, as well as section 6 of chapter 2, it should be clear that in the process of writing this memoir (over a period of approximately five years), we discovered some interesting problems for further research. Naturally, after such a long period of time, our hindsight has improved considerably. For example, if we had integrated by parts slightly differently in the display after (4.26) in chapter 1, we would have obtained

$$[p_1^m p_5]_{i,\eta}(\theta_\lambda^{(y_0,s_0)}) = m[p_1^{m-1} p_5]_{i+1,\hat\eta}(\theta_\lambda^{(y_0,s_0)})$$
$$+ \lambda^{-\alpha}[p_1^m p_5]_{i+1,\hat\eta}(c\phi_\lambda^{s_0}(\phi')_{\lambda^\alpha}^{y_0})$$

where $\hat\eta_{i+1} = (1,+)$ and $\eta_j = \hat\eta_j$, otherwise. L^∞ estimates on the second operator on the righthand side of this equation are easily obtained since the kernel of this operator has no singularities. Thus from the induction scheme in section 4 of chapter 1 we would only have needed to consider the first operator on the right hand side of this equation. Tracing the argument further one sees that in order to prove Theorem 1 of chapter 1 it would have sufficed to consider only those operators in section 3 of chapter 1 for which $\eta \equiv \langle(1,+)\rangle$. Now these operators are much easier to handle notationally and they have anti-symmetric kernels in the space variable, so it is easy to show they satisfy the hypotheses of the $T1$ theorem. We hope, though, that boundedness of the family of operators considered in section 3 of chapter 1, is of some interest in itself .

Also we note that if we were to write this memoir today we would perhaps consider even more general singular integral operators than those in Theorem 1 of chapter 1. For example, as pointed out in section 6 of chapter 2, one needs to obtain bounds for slightly more general operators, in order to obtain parabolic analogues of some theorems of David on regular surfaces. Also in conversations with Steve Hofmann (oral), the authors discovered that the estimates and iteration scheme in chapters 1 and 2 could be used to obtain higher dimensional analogues of the half derivative problem in space. Thus it would be tempting (although perhaps more difficult for the reader to follow) to write an all encompassing Theorem 1.

For a proof which includes all these features and handles all of the above multilinear singular integrals in a very unified way see [H1]. Also in [H1] an ingenious direct proof is given of the boundedness of the double layer heat potential on L^p as well as a proof that the single layer heat potential of an L^p function has 1/2 derivative in the time variable in L^p.

Next we point out that the authors discovered again in conversations with Steve Hofmann that conditions (1.1), (1.4), and (1.7) in this paper, taken together, are actually equivalent to a multiplier condition in [H], as well as condition (1.1), (1.4), and (1.15) in chapter 2. The equivalence of these conditions can be shown more or less directly or by using the $T1$ theorem to show that certain operators are bounded on $L^2(\ R^n\ )$. As regards the latter method, it follows from (6.28) in section 6 that if f satisfies (1.7), then f satisfies (1.15) (assuming (1.1) and (1.4)). To prove the reverse statement let K be the operator defined for $g \in L^2(\ R^n\ )$ by

$$Kg(x,t) = \int_{R^n} \frac{(f(y,t) - f(y,s))^2}{(s-t)^{(n+3)/2}} \exp\left[-\frac{|x-y|^2}{|s-t|}\right] g(y,s)\,dyds.$$

Using the parabolic version of the $T1$ theorem given in section 1 of chapter 1 and (1.1), (1.4), (1.7), it is not difficult to show that K is bounded on $L^2(\ R^n\ )$. In fact (1.7) , (1.4), (1.1) are equivalent to the statement that $K(\chi_{Q^*})$ is in BMO(R^n) on rectangles whenever Q^* is a rectangle in R^n . Now using the second part of the

David Buildup Scheme as in chapter 2, Lemma 4.2, it can be shown that K remains bounded on $L^2(R^n)$ under assumption (1.15) of chapter 2. From the equivalence stated above we conclude that the two conditions are equivalent. The equivalence of (1.7) and Hofmann's condition can be proved by showing that (1.1), (1.4), and (1.7) are necessary and sufficient for the L^2 boundedness of the first commutator considered in [H], since Hofmann's condition is also necessary and sufficient. Again a proof is easily given using a parabolic version of the $T1$ theorem.

8. References

[B1] R.M. Brown, *The method of layer potentials for the heat equation in Lipschitz cylinders,* Amer. J. Math. **111** (1989), 359-379.

[B2] R.M. Brown, *Area integral estimates for caloric functions*, Trans. Amer. Math. Soc. **315** (1989), 565 - 589.

[CF] R. Coifman and C. Fefferman, *Weighted norm inequalities for maximal functions and singular integrals*, Studia Math. **51** (1974), 241-250.

[FGS] E. B. Fabes, N. Garofalo and S. Salsa, *Comparison theorems for temperatures in noncylindrical domains,* Atti Accad. Naz. Lincei, Rend. Cl. Sci. Fis., Mat. Nat. **8-77** (1984), 1-12.

[FR] E.B. Fabes and N.M. Riviére, *Dirichlet and Neumann problems for the heat equation in C^1 cylinders,* Proc. of Symposia in Pure Math. (AMS) **35** (1979), part 2, 179-196.

[H] S. Hofmann, *A characterization of commutators of parabolic singular integrals,* to appear.

[H1] S. Hofmann, *Multilinear singular integrals, rough operators, and caloric layer potentials,* to appear.

[JK] D. Jerison and C. Kenig, *Boundary behavior of harmonic functions in non-tangentially accessible domains,* Adv. in Math. **47**(1982), 80-147.

[K] J.T. Kemper, *Temperatures in several variables: kernel functions, representations, and parabolic boundary values,* Trans. Amer. Math. Soc. **167**(1972), 243-262.

[LM] J.L. Lewis and M.A.M. Murray, *Absolute Continuity of Parabolic Measure*, in Partial Differential Equations with Minimal Smoothness and Applications, IMA Volumes In Mathematics And Its Applications **42** (1992), Springer-Verlag, 173-189.

[LM1] J.L. Lewis and M.A.M. Murray, *Regularity Properties of Commutators and Layer Potentials Associated to the Heat Equation*, Trans. Amer. Math. Soc. **328** (1991), 815-842.

[LS] J.L. Lewis and J. Silver, *Parabolic measure and the Dirichlet problem for the heat equation in two dimensions,* Indiana U. Math. J. **37** (1988), 801-839.

[S] R.S. Strichartz, *Bounded mean oscillation and Sobolev spaces,* Indiana U. Math. J. **29** (1980), 539-558.

[T] A. Torchinsky, *Real variable methods in harmonic analysis*, Academic Press,

1986.

[W] J.M. Wu, *On parabolic measures and subparabolic functions*, Trans. Amer. Math. Soc. **251** (1979), 171-185.

Department of Mathematics, University of Kentucky, Lexington, Kentucky, 40506 - 0027
email : john@ms.uky.edu

Department of Mathematics, Virginia Polytechnic Institute, and State University, Blacksburg, VA 24061-0123
email: murray@math.vt.edu

Editorial Information

To be published in the *Memoirs*, a paper must be correct, new, nontrivial, and significant. Further, it must be well written and of interest to a substantial number of mathematicians. Piecemeal results, such as an inconclusive step toward an unproved major theorem or a minor variation on a known result, are in general not acceptable for publication. *Transactions* Editors shall solicit and encourage publication of worthy papers. Papers appearing in *Memoirs* are generally longer than those appearing in *Transactions* with which it shares an editorial committee.

As of December 7, 1994, the backlog for this journal was approximately 3 volumes. This estimate is the result of dividing the number of manuscripts for this journal in the Providence office that have not yet gone to the printer on the above date by the average number of monographs per volume over the previous twelve months, reduced by the number of issues published in four months (the time necessary for preparing an issue for the printer). (There are 6 volumes per year, each containing at least 4 numbers.)

A Copyright Transfer Agreement is required before a paper will be published in this journal. By submitting a paper to this journal, authors certify that the manuscript has not been submitted to nor is it under consideration for publication by another journal, conference proceedings, or similar publication.

Information for Authors and Editors

Memoirs are printed by photo-offset from camera copy fully prepared by the author. This means that the finished book will look exactly like the copy submitted.

The paper must contain a *descriptive title* and an *abstract* that summarizes the article in language suitable for workers in the general field (algebra, analysis, etc.). The *descriptive title* should be short, but informative; useless or vague phrases such as "some remarks about" or "concerning" should be avoided. The *abstract* should be at least one complete sentence, and at most 300 words. Included with the footnotes to the paper, there should be the 1991 *Mathematics Subject Classification* representing the primary and secondary subjects of the article. This may be followed by a list of *key words and phrases* describing the subject matter of the article and taken from it. A list of the numbers may be found in the annual index of *Mathematical Reviews*, published with the December issue starting in 1990, as well as from the electronic service e-MATH [**telnet e-MATH.ams.org** (or **telnet 130.44.1.100**). Login and password are **e-math**]. For journal abbreviations used in bibliographies, see the list of serials in the latest *Mathematical Reviews* annual index. When the manuscript is submitted, authors should supply the editor with electronic addresses if available. These will be printed after the postal address at the end of each article.

Electronically prepared manuscripts. The AMS encourages submission of electronically prepared manuscripts in $\mathcal{A}\mathcal{M}\mathcal{S}$-TeX or $\mathcal{A}\mathcal{M}\mathcal{S}$-LaTeX because properly prepared electronic manuscripts save the author proofreading time and move more quickly through the production process. To this end, the Society has prepared "preprint" style files, specifically the amsppt style of $\mathcal{A}\mathcal{M}\mathcal{S}$-TeX and the amsart style of $\mathcal{A}\mathcal{M}\mathcal{S}$-LaTeX, which will simplify the work of authors and of the production staff. Those authors who make use of these style files from the beginning of the writing process will further reduce their own effort. Electronically submitted manuscripts prepared in plain TEX or LATEX do not mesh properly with the AMS production systems and cannot, therefore, realize the same kind of expedited processing. Users of plain TEX should have little difficulty learning AMS-TEX, and LATEX users will find that AMS-LATEX is the same as LATEX with additional commands to simplify the typesetting of mathematics.

Guidelines for Preparing Electronic Manuscripts provides additional assistance and is available for use with either AMS-TEX or AMS-LATEX. Authors with FTP access may obtain *Guidelines* from the Society's Internet node `e-MATH.ams.org` (130.44.1.100). For those without FTP access *Guidelines* can be obtained free of charge from the e-mail address `guide-elec@math.ams.org` (Internet) or from the Customer Services Department, American Mathematical Society, P.O. Box 6248, Providence, RI 02940-6248. When requesting *Guidelines*, please specify which version you want.

At the time of submission, authors should indicate if the paper has been prepared using AMS-TEX or AMS-LATEX. The *Manual for Authors of Mathematical Papers* should be consulted for symbols and style conventions. The *Manual* may be obtained free of charge from the e-mail address `cust-serv@math.ams.org` or from the Customer Services Department, American Mathematical Society, P.O. Box 6248, Providence, RI 02940-6248. The Providence office should be supplied with a manuscript that corresponds to the electronic file being submitted.

Electronic manuscripts should be sent to the Providence office immediately after the paper has been accepted for publication. They can be sent via e-mail to `pub-submit@math.ams.org` (Internet) or on diskettes to the Publications Department, American Mathematical Society, P. O. Box 6248, Providence, RI 02940-6248. When submitting electronic manuscripts please be sure to include a message indicating in which publication the paper has been accepted.

Two copies of the paper should be sent directly to the appropriate Editor and the author should keep one copy. The *Guide for Authors of Memoirs* gives detailed information on preparing papers for *Memoirs* and may be obtained free of charge from the Editorial Department, American Mathematical Society, P. O. Box 6248, Providence, RI 02940-6248. For papers not prepared electronically, model paper may also be obtained free of charge from the Editorial Department.

Any inquiries concerning a paper that has been accepted for publication should be sent directly to the Editorial Department, American Mathematical Society, P. O. Box 6248, Providence, RI 02940-6248.

Editors

This journal is designed particularly for long research papers (and groups of cognate papers) in pure and applied mathematics. Papers intended for publication in the *Memoirs* should be addressed to one of the following editors:

Ordinary differential equations, partial differential equations, and applied mathematics to JOHN MALLET-PARET, Division of Applied Mathematics, Brown University, Providence, RI 02912-9000; e-mail: am438000@brownvm.brown.edu.

Harmonic analysis, representation theory, and Lie theory to ROBERT J. STANTON, Department of Mathematics, The Ohio State University, 231 West 18th Avenue, Columbus, OH 43210-1174; electronic mail: stanton@function.mps.ohio-state.edu.

Ergodic theory, dynamical systems, and abstract analysis to DANIEL J. RUDOLPH, Department of Mathematics, University of Maryland, College Park, MD 20742; e-mail: djr@math.umd.edu.

Real and harmonic analysis to DAVID JERISON, Department of Mathematics, MIT, Rm 2–180, Cambridge, MA 02139; e-mail: jerison@math.mit.edu.

Algebra and algebraic geometry to EFIM ZELMANOV, Department of Mathematics, University of Wisconsin, 480 Lincoln Drive, Madison, WI 53706-1388; e-mail: zelmanov@math.wisc.edu

Algebraic topology and differential topology to MARK MAHOWALD, Department of Mathematics, Northwestern University, 2033 Sheridan Road, Evanston, IL 60208-2730; e-mail: mark@math.nwu.edu.

Global analysis and differential geometry to ROBERT L. BRYANT, Department of Mathematics, Duke University, Durham, NC 27706-7706; e-mail: bryant@math.duke.edu.

Probability and statistics to RICHARD DURRETT, Department of Mathematics, Cornell University, White Hall, Ithaca, NY 14853-7901; e-mail: rtd@cornella.cit.cornell.edu.

Combinatorics and Lie theory to PHILIP J. HANLON, Department of Mathematics, University of Michigan, Ann Arbor, MI 48109-1003; e-mail: phil.hanlon@math.lsa.umich.edu.

Logic and universal algebra to GREGORY L. CHERLIN, Department of Mathematics, Rutgers University, Hill Center, Busch Campus, New Brunswick, NJ 08903; e-mail: cherlin@math.rutgers.edu.

Algebraic number theory, analytic number theory, and automorphic forms to WEN-CHING WINNIE LI, Department of Mathematics, Pennsylvania State University, University Park, PA 16802-6401; e-mail: wli@math.psu.edu.

Complex analysis and nonlinear partial differential equations to SUN-YUNG A. CHANG, Department of Mathematics, University of California at Los Angeles, Los Angeles, CA 90024-1555; e-mail: chang@math.ucla.edu.

All other communications to the editors should be addressed to the Managing Editor, PETER SHALEN, Department of Mathematics, Statistics, and Computer Science, University of Illinois at Chicago, Chicago, IL 60680; e-mail: shalen@math.uic.edu.